Publish and be damned
www.pabd.com

An Indian Calling

Mick Sheridan

Publish and be damned
www.pabd.com

© Mick Sheridan

This work is licensed under the Creative Commons Attribution License. To view a copy of this license, visit http://creativecommons.org/licenses/by/2.5/ or send a letter to Creative Commons, 559 Nathan Abbott Way, Stanford, California 94305, USA.

First published in Great Britain 2005 by Mick Sheridan
The moral right of Mick Sheridan to be identified as the author of this work has been asserted.

Designed in London, Great Britain, by Adlibbed Limited.
Printed and bound in the UK by 4Edge.

ISBN: 1-905452-20-9

This work is part fact, part fiction, where facts have been embellished, names have been changed to protect the innocent

Publish and be Damned helps writers publish their books with absolutely no set-up costs or publishing fees. The service uses a series of automated tools to design and print your book on-demand, eliminating the need for large print runs and inventory costs. Now, there's nothing stopping you getting your book into print quickly and easily.

For more information or to visit our book store please visit us at www.pabd.com

Chapter One

And I quote:

'You know what you can do with your job? You can take it, and your eye teeth, and your franking machine, and all that other rubbish I have to go about with, and you can shove it right up your arse'.

Of course it wasn't me who said that, it was young James Cooper, the working class hero in Quadrophenia. I didn't say it because I'm not Jimmy - I'm not nineteen and I'm not working class, at least not any more. At thirty five I still do rather see myself as a mod, obscene though that may be, but I don't have the nervous hatred any more, I don't have the bile, even if I do have the parka. So although I longed to say what Jimmy said, my civility and judiciousness required that I instead tendered:

"There's no easy way to say this boss, so I'll come right out with it. Something has come up in my personal life which means I'm going to have to leave the company."

Which does not have quite the same impact you'll agree, and I buffered it further to extinguish any potential burning bridges.

"Well it's Julie really, she wants to spend a year or so in Goa, India, and I've agreed to go with her. I'm not going to work, I'm going to write a book."

I was too spineless to take the blame myself so I did what so many men do when they lack the courage to deny their peers and blamed the wife. And of course my resignation was quickly and professionally accepted with a well wishing handshake. No imploring or beseeching, not even any real expression of sorrow, just another businesslike anticlimax.

I wish resigning from a job could be a euphoric, defining moment; a defiant gesture of independence; two fingers to the system, like it is for Jimmy. But it never is, at least not for me. The problem is I think I'm more important than I am, and I expect the company to be devastated at the loss. But, like Jimmy, I will simply be replaced by someone else, and the business will resume largely unaffected.

Up to now, my line of work has been market research. A career I had not planned, but one that had already spanned some fifteen years. It began by accident and had continued by accident. I was in charge of a call centre conducting market research, hiring people on a freelance basis to phone other people to answer questions on a good-will basis. I largely employed layabouts and drifters who needed short-term money without having to sell out long term, and I liked this because I could continue to imagine that I had not sold out either. But the devious commercial spirit surreptitiously possessed me, killing the inner Trotsky, and after less than five years, I was a corporate bore, taking it all too seriously, seeing it symbiotically; the job and I depended on each other. Like so many others, I had been drugged by the comfort of a regular salary, institutionalised by the money - the salary as pacifier.

I had been trying to justify this concept of the salary as pacifier to myself for some years, trying to substantiate money as medicine, seeking to mitigate my guilt for becoming a bread head. But I'd failed and could only claim diminished responsibility as a corporate junkie. I am frankly amazed at how easy it is to be allayed by promotion - a pat on the back and a small hike in income is all that I've needed to follow doggedly in *the way of the work*. There is nothing as flattering as being worth more money. Each time I had become disillusioned in a job, my wily superiors seized the opportunity and gave me a higher dose of the pacification soma. And I lapped it up. This was not just the work treadmill any more, not the trap of paid labour versus poverty that is quite easy to beat for us would-be Bohemian types. No, this is psychological containment, the Catholic work ethic perhaps: guilt, money and obedience - the comfort of institutionalised security. They give it to us and we take it, we *enjoy* it, and who wouldn't? It is *enabling*. The money has enabled me to *take part* in the conversations about house prices. It has enabled me to travel around the world on sanitised two-week package breaks to meet other well-heeled salary slaves to discuss house prices and the lack of Health & Safety in the developing world.

It took fifteen years to recognise I'd been duped. A customary mid-life crisis meant it had to change, and I decided I was to be pacified no more; from here on I would live without corporate sponsorship, without pandering to the pay rise, without the buffering bonus. I had to leave the country to be sure of escape. I'd go somewhere where I could not be tempted by money or a better job, somewhere that my company skills were useless. I would go to India.

Cold Turkey. No rehab for me, just straight up No Work, for a year at least. I was sure I could break the cycle and get off the salary for good, but I went to India to be sure. India, a place of spirituality and sense, a place of frequent madness and enormous frustration, the one place I knew where business couldn't tempt me, where pay was so bad that I could ignore its lure, where my skills as a call centre manager would be useless, I mean, they don't have call centres in *India*.

Hang on a minute...

Chapter Two

"That will be one hundred and twenty five pounds please."
The luggage was over weight. Only twenty kilogrammes per person came free with the ticket. We were emigrating for Christ's sake, we couldn't emigrate with just a knapsack. Julie and I were taking this seriously, we weren't *tourists*, we were starting a new life in a new land, and an extreme land. We'd need a bit more than a few clothes, a penknife and a guidebook.
"One hundred and twenty five pounds please."
I was already dog-tired from the 4am start and the hefty travel itinerary was fast becoming a demanding monkey on my grumpy back. I mumbled a stream of unconsciousness at the excess baggage desk, "cheapo airline", "British Airways next time", "You're all bastards", that sort of thing, and as I handed over a credit card I stiffly pooh-poohed the clerk, shooting the messenger. I turned to Julie, "It's started."
The weeks preceding departure bulged with activity. Preparations aside, the most wearing of all were the goodbye ceremonies. First work, then family, then friends, then the friends who couldn't make it, then the 'close' family, and finally the last night. Each time an indecent amount of emotion and each time a decent hangover. The emotional mix played havoc with my equilibrium, culminating in a succession of melodramatic outbursts and denials of the 'Oh God, this is all too much, I just can't go, we'll have to cancel' ilk. Fortunately for me, Julie is a strong character and weathered it all without complaint.
We had previously been to India a few times and we felt we knew roughly what to expect and how to handle it, but the enormity of a relocation to the developing world should not be underestimated. What does one *pack* for such a transfer?
Our prior experience had been one-month-at-a-time holidays and only ever during the British winter. We had always been sun-seeking tourists and had very little notion of what a full year of Indian weather would bring. We of course knew that after the very hot summer there would be a very wet monsoon, so the first thing into the case was the umbrella, raincoat and sturdy walking boots. Next came the numerous shorts,

T-shirts, swimming stuff, hats, sandals and sun care products that we would need for the hot periods, quickly followed by the long trousers, jumpers and jackets we would obviously need when the weather started to turn cooler. Then came the products; because of course, shaving gel and razors wouldn't be available in India, and presumably, neither would drugs, books, HP sauce, art materials, tea bags, Marmite, toothpaste, cotton buds or a skipping rope.

I stood back and looked at the two coffin sized trunks we had stuffed and felt like William Boot from Evelyn Waugh's 'Scoop', minus the cleft sticks. Once I'd added the laptop, Gameboy, Walkmen, some 200+ Cds, camera equipment, minidisc recorder and football memorabilia, I felt like William Boot with the cleft sticks *and* the canoe.

We were emmigratory dunces with the dumb English notion that the world was our birthright, that this momentous change to where we live and to our lives, our rite of passage, was instead our Right of Passage. We were from the old 'pink section' of the map and were porting our pink skins to where our belligerent colonial ancestors had so brazenly planted their flags and their tea. To the land of the Great British Curry, the Morris Oxford and the Enfield Bullet. We had done next to no research and were in all likelihood, woefully under prepared. We had chiefly chosen India because it was hot and cheap. We reckoned we could afford to live lacking the security of a wage but without venturing into the dangerous insecurity of debt. But we were not simply tight-arse travellers, roughing our way round the subcontinent, existing on rice and *dal* and a once-a-week beer. No sir. The backpacker fraternity had nothing on us. We were at the vanguard of a whole new breed of wageless wayfarer: the Nouveaux-Skint. True, we had a bank account stuffed with cash, but no way were we going to spend it. Downsizing, downshifting and downright stubbornly not going to spend our money. This was something of a departure for us. We are 'sensible' spenders but ones who cannot resist the temptation to be flash. Our membership of the New Tight must begin with a budget emigration programme, which must begin with a budget airline.

The gulf between real budget travel and, say, British Airways, is chasmic. Budget travel to India exemplifies the divide. Our chosen route was via Damascus, where we hoped most of the passengers would disembark. We had assumed that Syrian Arab Airlines would be full of Syrians and Arabs - how wrong we were. As we took our seats I realised an ordeal had begun, and a very Indian ordeal. The tatty, unvaleted interior carried a vague odour of old sweat with a lingering of old Old Spice; but omnipresent was the unmistakable aroma of unrest. The plane was already fully rammed with Indians and their belongings. All aisles and exits were blocked and there was mounting pressure from the bulging queue behind me. It felt like an Indian microcosm already, and I noted again how every Indian scene seems to contain humans, and usually in large numbers and positioned in very close proximity. Even those panoramic shots of wild and mountainous Kashmir - zoom in close and there's always at least one man there half way up the mountain or standing at the edge of the lake, I have no idea what he is doing, but he is definitely there. For a country with over a billion inhabitants this should come as no surprise, but somehow I'm never fully prepared for it. I am English, and as such carry a stuffy disposition of aloofness around with me. Furthermore, I value personal space so highly as to be considered by many as unfriendly. In India this attitude gets you nowhere, or more accurately, it gets me nowhere. In my experience Indians are, in the main, happy, friendly, tactile people, relaxed and at ease with themselves who, in a travelling situation, will always talk to their neighbour, help them with their luggage and generally be nice. However, in a long-haul flight situation this will almost always progress to removing footwear, *spreading out and relaxing* directly onto their neighbour and very often culminate in full on, mouth open, flip top head shout-snoring.

Being of a somewhat fractious and positively short-tempered disposition, I approach such situations in a pessimistic state of resigned doom, and my current circumstance proved no exception. With a gentlemanly foreboding I ushered Julie in to her window seat and attempted to place our bulging hand luggage in the overhead lockers. Full. With great gusto and without due care or attention, my bags were seized by my smiling, across-aisle neighbour and stuffed into the lockers about four rows back. I thanked him for his kindness through tight lips and took my central seat in stiff reservation to await my neighbour-fate.

Normally I am not the first passenger to board a plane, in fact I usually try to be the last. As a childish game I enjoy sitting back and nonchalantly watching a melee ensue as passengers are invited to board. I have no idea what makes people rush for the door in this way or indeed, why would anyone would want to stand in a queue when seat allocation is prescribed. I love to sit and mock them as they scramble, and relish in being the last one sitting. There are usually about three contenders for the accolade of 'last one on', and I am usually among them. It's a case of who will be the first to crack as we sit reading, pretending not to have noticed the aforementioned scrum and prevailing lack of it. In truth I always become slightly nervous as the hostess starts to square up the boarding cards, tapping them on the table and looking impatiently at the three stoic stragglers, but I try not to show it and I often win. It reminds me of trying to be the last one to clap in school assembly, holding out a full second after the last clap then hearing one come in immediately afterward, and knowing that any more will bring punishment. I am not a very competitive man, I have no sporting trophies on my mantlepiece, but these challenges are relish. But here I felt different, I felt the need to stamp some authority on things, to get to my seat and stake my claim, it would be a long flight and I wanted to make it as comfortable as possible. I boarded boldly and rudely, ignoring the singsong welcomes of the air hosts and hostesses, surging quickly forward to find my seat. Sitting next to Julie on one side and an empty seat on the other, waiting to see with what horror I would have to rub shoulders all night, I experienced a most unexpected moment of optimism, what if the seat remained empty?

It was short lived. As I saw the hippopotamus bumbling down the aisle towards me, I knew...

After seven full breathless hours we reached Damascus. For me the Road to Damascus had been a very long, thin, squashed road. In fact my neighbour had been a nice man. He hadn't removed his shoes, didn't snore, was congenial in conversation and graceful when asked to stand while I passed to go to the toilet. He was, however, a man of *weight*, and this was an aircraft of *economy*, a merciless combination. But for me he had one great advantage over most of the other passengers - he was Syrian, and he was getting off. We bid each other farewell and my body resized as he rose, like a cartoon character popping back into shape. I looked over at my across-aisle neighbour and glowered as he eyed the empty seat, I caught his eye and growled audibly. He looked at me imploringly and I noticed that he too had a fat neighbour, and worse, an Indian who would be bulging all the way to Bombay. How fate has a cruel hand.

Next stop Abu Dhabi. No one seemed to board the plane at Damascus, we certainly had fewer passengers now, but the noise level had risen in steady defiance. Dinner was served; a small piece of stinking lamb accompanied by five shrunken, over-stewed green beans and two par-mashed hexagonal carrot slices, served up by our hassled, hustled low-rent hosts and in the main shunned by the largely vegetarian clientele. It seemed the majority of passengers had neglected to pre-book their vegetarian meal, and those who had were being palmed off with a meaty option anyway. One sure way to insult a vegetarian is to serve up meat. Offence was taken causing further clamour and clatter; a wrong-dinner din. I pushed the food around the little tin while everyone shouted at each other, wondering whether to chance eating it. I have a cast iron stomach that allows me to eat anything without fear of repercussion, so I ate, and as always was strangely satiated by the poor quality of the fare. I fear I am quite alone in an unlikely fondness for poor airline food, so will dwell on it no longer lest my isolation intensifies.

An hour refuelling in Bahrain and I was dying to disembark. The lack of legroom had by now become the major irritant, the restlessness of fellow passengers being darkly overshadowed by my own bad tempered legs. They had begun to twitch, fizz, and throb in that way, which seems exclusive to confined travel; the bones began to vibrate, resonate even, like tuning forks inside my legs. Standing up offered very short-lived respite, and the act of compressing myself back in to the sardine can seat once again served only to acerbate my disposition and intensify the leg throb. I thrashed in the seat like a new trout, occasionally spinning like a lassoed croc. I whinged and moaned at Julie. She remained aloof, leaving me to it. Finally we took off. The twitching, fizzing and throbbing persisted. There were three or four hours to go and a panic over the legs was fast approaching. Action needed. I decided to spend some time in the toilet. On my way I spotted an oasis - three empty seats in a row, a mirage surely? I blinked hard and approached with stealth. I sat in the aisle seat and turned to smile at my neighbours, looking for reassurance, seeking their approval. They were all asleep. I needed no further invitation and moved silently into the central seat, pushing back both armrests and pivoting on my arse-fulcrum with ease. Delight. I silently filled the three-seat space, kicked off my unlaced shoes, and felt calm spread through my lower body like love.

As the plane descended, the familiar Indian scenes unfolded on the ground before us. The city slums or *wastis* were the first visual indicator that we were arriving unmistakably in India. The tightly huddled 'hutments' of mud, corrugated iron, burlap and tarpaulin plug every gap of city space, the effluent mortar around high rise bricks. In such slum settlements, a single outhouse may be shared by literally thousands of people, or, more usually, there are no sanitary facilities at all. Ditches are awash in raw sewage, and byways are strewn with the refuse of people and animals with nowhere else to go. In many sections of Indian cities, scavenging pigs, often owned by *sweepers*, along with stray dogs, help to recycle fecal material. Piles of less noxious vegetal and paper garbage are sorted through by the poorest people, who seek usable or saleable bits of things. Cattle and goats, owned by entrepreneurial folk, graze on these piles, turning otherwise useless garbage into valuable milk, dung (used for cooking fuel), and meat.

By contrast, downtown Mumbai is predominantly peopled by the conspicuous growing ranks of the middle class. Consumer brands are everywhere and today's Indian yuppie is every part of his 1980's counterpart in the West, the moustache perhaps notwithstanding. A sea view apartment on Marine Drive will set you back far more of your monthly salary at the stock exchange than a Thames view ever did, but the Chevrolet or Skoda you drive will cost a good deal less than your Golf GTi.

Poverty is so prevalent in India that one becomes unsentimental about it in a surprisingly short space of time. I recall the first time I saw naked children sprawled in the shit and piss on the street and how I sat alone and cried for the world. But I soon toughened up. Before I knew it I was waltzing past the beggars with a raised palm, ignoring even the most piteous plea for alms, having become desensitised by the ubiquity of the deprived. In my defence I can offer very little, except perhaps to say, as Mark Tully (the BBC's Asia Correspondent for so many years) does, that 'I'm not alone in this: most prosperous Indians - and indeed the prosperous in all parts of the world - have learnt to live with the fact that millions of Indians live below what economists define as the poverty line.' I have succumbed and have myself adopted Tully's response to the question everyone asks, "How do you cope with the poverty?" by retorting simply and sadly "I don't have to. The poor do."

We finally landed in Mumbai. I never know whether to call it Mumbai or Bombay. Since the name change in 1995 which was as much to do with regionalist language politics as it was with anti colonialist feeling, it has become a dual city whose denizens might refer to themselves as both Bombayites and Mumbaikers in the same sentence. Like so many other amendments that Indians have been forced to make to their lives throughout history, the change was both accepted and contested, endorsed by some, refuted by others, but used to some degree by all. Constantly different and always the same. 'Same, same'.

Stepping out of the aircraft, the smell and the heat punched me in the face and I grabbed the handrail to steady myself before descending carefully to the scorching tarmac. I smiled a big smile, the journey wasn't quite over, but the worst was, and as they say in India, ' We'd reached.'

We had booked a connecting flight to Goa but this would not be available for check-in for a further five hours. The free shuttle service would take us from the International Terminal to the Domestic Terminal, a distance of two hundred metres. We waited for it while others grew impatient and succumbed to the bellowed offers of cheap taxis. One hundred rupees for a two-minute journey; we knew it was a rip off because my impatience had led me to fall for it before. The first time I came here I was terrified, but now I felt good, I felt in charge. In the main terminal we ate and drowsiness prevailed. We'd been travelling some twenty-eight hours by this time and sleep was imminent, regardless of whereabouts. I am not a public sleeper, in fact I generally cannot sleep unless I am in a bed, but I leant on my luggage and slept hard and (probably) loud. I was woken by a small, irritable, animated man in military uniform, tugging at my supine elbow; I drew my head from the luggage-pillow and slurped a large dribble up one side of my mouth.

"What? What is it? What's wrong"

"Sir, Sir, Goa flight sir?"

"Yes"

"Quick Sir, Quick."

We checked in with only minutes to spare in a state of goggle-eyed delirium. As we approached the security check I was *wagging* furiously - a state somewhere between sleep and furiousness. I put some loose change, the only metal objects about my person, in the plastic tray and walked through the security scanner. It gave the inevitable beep. I am not metal, but I always beep. I turned out my pockets further revealing nothing. I was circumnavigated with a hand-held scanning device by a very angry man who was determined, but found nothing. I went back through the scanner, no beep this time. I was waved on.

Many transactions in India are needlessly bureaucratic. Airport check in and arrival are astonishingly so. On one occasion leaving Kerala I went through eleven different procedures with eleven different individuals before boarding the plane. These were: 1. Show ticket to security guard to enter airport. 2. Give ticket to seat allocation desk. 3. Give luggage to separate check-in desk. 4. Collect boarding card from another desk. 5. Identify luggage to porter for loading. 6. Security check one. 7. Passport control and immigration. 8. Security check two. 9. Redeem part of boarding card to enter runway. 10. Identify luggage again on runway. 11. Walk to and board the only plane on the runway. Possibly this is all due to India's commitment to the provision of full employment, but more likely it is due to straightforward lunacy.

After the inevitable beeping episode, I approached Security Check Two in a sluggish drowse, aware that I would not be free from bureaucratic interference for quite a stretch. I was summoned by a sombre female to park my hand luggage on the table before her for inspection. As I did so I gave myself quite a stretch and quite a yawn, apologising for my tiredness and trying to be good-natured through idle conversation. But she was having none of it. I had picked the bastard, the overly fastidious, bitterly suspicious, pedant-bitch. She emptied everything out. She opened my laptop computer and said 'Computer', then progressed through my belongings, naming each one in turn in some banal Generation Game performance, or perhaps she was auditioning for Catchphrase - 'Say What You See'. She opened a large case containing CDs and said 'CDs'.

"Yes, around two hundred of them I shouldn't wonder, they're amazing, when you put them in this player here (holding it aloft) they play music. It's like magic. You wouldn't believe it. Absolute magic."

Registering my irritation, she proceeded to look behind the CDs, into each of the sleeves of the CD wallet. I sighed, knowing I had brought it on myself. Eventually she brushed my belongings to one side and beckoned the next passenger, leaving me to huff and puff my things back into my bag. As I walked toward the next official in resigned somnolent silence, I realised I had left the loose change in the plastic tray by the scanner. It was only around fifty pence in Sterling but I was irritated and went back for it. It wasn't there. I approached the guard.

"I left my money in this tray, where is it now?"

He eyed me with disinterest and turned away. I repeated myself in a raised tone and was again ignored. I started to colour and shake a little, visibly losing my temper; another guard stepped forward. He stooped toward me and pulled some coins from his pocket, it was my money. He handed it to me.

"You were going to keep that weren't you? Surely that is theft? I was only standing over there being harassed by that she-devil and you stole it right under my eyes."

He rapped his *lathi* hard on the table next to me. "Why did you leave this money here? Why did you not take it with you?"

I was nonplussed. I had not anticipated violent inquisition. I simply stared at him, mystified.

"Well", he roared, "Why did you leave it?"

I felt weak. My head clouded and I fought for an answer. "It was a mistake, I didn't leave it there on purpose for God's sake, I don't leave money behind on purpose."

He barked a military retort "Don't leave money here again", and strode away.

I turned and strode off, angry about this injustice, but as I reached the departure lounge with the silver in my hand, an unexpected smile began to spread slowly across my face. As I approached Julie I was nodding to myself and noticeably grinning as I slowly talked myself back through the incident and noted how effortlessly and brilliantly he had turned his crime into my punishment. The fact is, I thoroughly enjoyed being rendered speechless by my adversary, particularly by one so overwhelmingly belligerent as this chap, there seemed something so *just* about being belittled in this way, I almost certainly *deserve* it in a *karmic* sense, for I have dished out enough (often unwarranted) belittlings myself and where better to be on the receiving end of such *karmic* payback than in India?

We finally boarded the Goa plane. The security incident had strangely perked me up, and thus cheered, I set my eyes about my first Indian Airlines flight. I had never flown a domestic Indian flight before, and was pleased by the sari-clad hostesses, their offer of complimentary mango *Frooti* (a fruit drink) and the phrasing in the English language safety demonstration which stated that it is the duty of those passengers seated near the exit doors to help others in an emergency, 'inconvenience is regretted'. The flight was only forty minutes, but lunch was to be served. I calculated that there would be enough time to serve lunch but not enough to eat it. I tore the lid from the little tin to reveal an Indian Flag of food. Three stripes - orange, white and green. An orange coloured gravy with potatoes, white steamed rice, and a paint-green *palak paneer*. The white rice had a cashew at its centre; the National Banner with the nut as its spinning wheel crest. As I have mentioned, I actually *like* airline food, mostly I just enjoy the silliness of serving hot food in such tiny and varied portions, but this was really excellent; real culinary kitsch. What we Brits call 'curry' lends very well to being put into a tin, and the food was quite edible. To arrange it into a national symbol is so perfectly gratuitous, and I would add, that I became a lifelong Indian Airlines customer there and then.

India has two main food distinctions: Veg and Non-Veg. Veg is fairly simple and refers to everything that is not meat. Non-Veg is also quite simple in that it refers to food that may or may not contain meat. This distinction is used in all situations (ok, there are variations like Pure-Veg which is without eggs and Eggetarian which is just a stupid thing altogether) and you can expect to be asked 'Veg or Non-Veg?' by many waiters so that they may give you the correct menu. I'm sure you get it by now: Veg = vegetarian, Non-Veg = meat .

Non-Veg is I think unique in the food world in being described as what it is not. In England, we do not have *non beans on toast* or *non boiled beef and carrots*. But in India one is very likely to encounter *non veg platter* and *non veg curry*. Personally I don't care what the food is or isn't, as long as it is wonderful, which it always is; the best cuisine in the world bar none (or maybe that's Non).

So after our kitsch dish, it was with renewed fervour that we touched down at our chosen destination. The hell of the mammoth journey was surpassed by the excitement of finally reaching our destination. A sense of nervous exhilaration and anticipation at what was to come; had we done the right thing? Where would we live? Could we settle? Are there really snakes?

****

We had chosen to live in Candolim in Goa. We had largely chosen this area because we knew a bar there that showed English football on TV and stayed open year-round, including monsoon. Looking back, this was pretty much the only reason we chose the area, a foolish whim perhaps, and certainly confirming our lack of plans. We had booked ourselves into the Hotel Casablanca in advance, a delightful little hostelry where we had stayed on a previous visit. We knew the owner and he was opening pre-season to accommodate us. Julie strode off to book a taxi while I lingered in trepidation at the luggage carousel. I have so far never lost checked-in luggage, but I know it is only a matter of time. As the conveyor trundled before me, I began to recall from bitter experience the condition I could expect our cases to be in. Indians have the worst luggage I've seen, worse even than the Irish. As a child I was shocked and almost embarrassed by the string bound suitcases and shameless blanket rolls carried on the Holyhead to Dun Laoghaire ferry. This was Irish working class-ness, an hereditary indifference to panache among those who would rather spend their hard earned money on a few pints than a new case. Times have changed, and the 1970's Pint of Guinness and 20 Major preference has gone. Ireland has changed its attitude to appearances and Habitat is selling well, so, probably, is Samsonite. But not in India. Samsonite is in fact not selling too well at all, despite attempts to promote it. Its main competitor, VIP, has an ubiquitous advertising campaign, and is probably fairing slightly better but neither company stand any real chance of success in the way that they have been able to in Ireland. India is different.

Traditionally, luggage has been about necessity in India. Luggage as a fashion item or statement is a relatively new concept, and one that may take a while to catch on. Portage is a means of thing-movement, bags are selected foremost for their strength, size and cost. Suitcases and trolleys are of sturdy economy and size is of far greater consideration than style. This does not mean that the average Indian is unconcerned with style, the issue is not one of a lack of chic, nor indeed, one of humility, rather it is an issue of necessity, even sanity. The fundament is that of luggage *handling*, in short, Indians cannot *handle* luggage, they treat it like shit.

The Indian traveller has no choice, wherever he goes his luggage is abused. At every opportunity, the *coolie* will throw luggage to the floor, smash cases against trunks, stuff bags into filthy spaces. All stowing facilities are indecent. Everybody involved in baggage handling delights in mistreating their quarry, biting the hand that feeds them. ACAB - all *coolies* are bastards. In its inaugural Indian journey, a beautifully crafted case will be reduced to a stinking heap of broken shit. A single flight will render each checked-in item totally and utterly ruined. And your Indian knows this. An Indian knows that if he spends good money on a lacquered flight case, it will be a scuffed, scratched, filthy cause of shame after its virgin voyage. Instead, the informed, resigned but discerning Indian buys cabin luggage that he can keep with him at all times. Everyone else simply gives in and buys disposable or terrible luggage: plastic bags; gunny sacks; cheap plastic suitcases made in Chinese sweatshops; hideously coloured canvas trolley cases from cheap market outlets; huge, nasty chintz-patterned trunks; and simply watches it deteriorate without anxiety. And who can blame them? To travel in India by any transport means, one must accept the fate of one's luggage. The happy traveller must resign himself to scruffiness in this respect. He must expect luggage ruination, for it will happen. He must be proud to arrive at his destination looking like an uncaring, unclean, unabashed piece of shit. Everyone else does.

Our luggage had been bought at a discount market stall and was most likely from their 'Death' range. That is, a range of products so cheap as to defy economics, the only possible explanation for which is that the manufacturer exploits the workforce so badly that many of them die during the process of manufacture, hence 'Death'. These products often bear well known brand names and I would normally avoid their purchase on conscientious grounds, however, on this occasion I could not ignore the beautiful irony in transporting my own largely Western items to the East in a suitcase with a small tin badge exclaiming 'Excellent' with a single internal label - 'Made in India'. I realised that this wonderful irony had backfired when the luggage finally made its way up the ramp and on to the conveyor. Both cases had broken handles and broken zips, one had ruined wheels, both were utterly filthy. As I wrestled their broken bulk from the carousel my mood darkened. I pushed my way towards Julie and the taxis, fighting off the insistent *coolies* and porters. Leaving any airport in India is always like running the gauntlet, and today this new anger fuelled my charge, like a borstal-boy rugby match I handed them off one by one on route for a try, and with small uniformed men strewn all over the field I reached the byline and our taxi.

Julie had chosen a quaint old 'Fiat' taxi, black with a yellow roof; a vehicle that would be described by most as having seen better days. Under normal circumstances, we both like this sort of car, for it has charm. Indeed, it is such charm that brings us to back this India so readily - this India of our hopes and dreams and holidays; this sun drenched Utopia where everyone is so affable and obliging, relaxed and pleasant, thoughtful and kind. But I was really too tired to be messing about with such whim after our marathon trek and longed for an air-conditioned vehicle for the final leg, so as I put the cases next to it and motioned toward the boot, I pulled a face at Julie that said, 'Why? Why? Why?'

"Pay first. Not this taxi, this man taxi." The driver screamed at us from his cab, pointing at an ancient, tiny, filthy man and his ancient, tiny, filthy taxi.

"I've already paid," Julie told me, presenting the pre-paid chit that she had somehow managed to procure while I was fussing with the cases, "But I think he means we have to go with this chap."

She was pointing at Lord Ancient Tiny Filth and seemed wholly unconcerned by the decrepit state of his vehicle, it seemed she was quite content to travel fifth class. I assented with a resigned glare and inspected a large area of rust around the near side suspension, thinking that I must check the condition of the engine mounts before we set off. The diminutive man manhandled our monumental luggage into his knackered little car. It would of course not fit.

"Right, we'll have to have another taxi," I stated in delight, striding off towards a lone, large HM Ambassador parked some twenty yards back in the taxi queue.

"Can you take us to Candolim please? Our luggage won't fit into the smaller taxis."

"I am not a taxi sir."

"I know *you* are not a taxi, but this fine specimen of post war British engineering surely is, it clearly states 'Tourist Vehicle' on the side. Now how much to Candolim my good man?"

"No, I am waiting for someone, I am booked sir."

I scanned the line further, and seeing no other large vehicles, I turned back to Julie, who, to my horror was helping Lord ATF to shove our cases on to a *roof rack*.

"I hope you're going to tie those down." I blurted.

"Look, stop panicking will you, this is fine, just get in, stop being such an old woman, I thought you like these cars anyway."

"Julie, this is no longer a car, it is…"

"Oh shut up and get in."

I acquiesced on the basis that if I did not, I would be considered an old woman. I sat in the front and looked scornfully at Lord ATF. I scanned the interior with distaste and settled back on the man. I glared at him with the most accusing and offensive face in my repertoire, folded my arms, sighed audibly, turned up my nose and looked away from him again.

He turned the key and the noise began. He beamed at me with his beatific smile, "Welcome to India Sir, welcome to India Madam. I will take you to Candolim, no problem." He beamed further as he engaged the first gear. I stewed in bile and hatred, offering him only rudeness.

As we pulled away from the airport I sat in silence and continued to show outward anger, but I was discomfited. I had behaved like a spoilt brat, and I was embarrassed about it. I stared out of the window in sulky protest as Julie chatted casually with the driver. As I began to take in the surroundings, I recalled the first time we had landed here, and what a culture shock it had all been. And how much I loved it.

I noted the smells again: the overriding thickness of the air in its mixed spice pungence; a masala of humidity, fire, sweet incense and flowers with the occasional waft of raw sewage to keep one's nose on its toes. The smell of India, and everything that is in it.

I ticked off the sights again: the bright sun glinting on the chrome bumpers of the Maruti Omni taxi vans rushing by roadside paddy fields tended by a spectrum of bending migrant labourers or *'dutties'*, people from the hills. The women perched on the Zuari bridge beside their small stacks of enmeshed green-backed crabs watching the roadside crows picking a rubbish heap apart, dancing around the humpy backed cows. The river bursting with monsoon rain, spilling into the red dirt streets to fuel the screaming bougainvillea flowers in pink and white. And adverts: Pepsi, Bisleri, Kingfisher, Cadbury, Kit Kat, painted on every possible edifice and precipice, every available construction, elevation and prefabrication.

I noted the sounds again: an endless cacophony of blaring horns, barking dogs, squawking crows, crying vendors, roaring engines and people hawking up great big spits, interspersed with fleeting impossible silence, the resonance of a soundless summer breeze anywhere. The sound of my new home.

My stupid swinging mood had swung back again and I now felt such a fool. The attack on the senses had done it. In a matter of fifteen minutes and a distance of around three miles, I had been altogether reminded of why we had come, of what we were doing here, and my childish anger had been eradicated. I now wanted to enjoy the experience but was stuck in a sulk, perpetually playing the teenager. I know of only one way to redeem oneself after a tantrum, and that is to publicly apologise. I sat up in my seat, and raising my shoulders and lifting my head high, I offered:

"I would like to wholeheartedly apologise to you both for my earlier rudeness. I have no reasonable excuse and look for no forgiveness, I will wear my embarrassment and stupidity as a badge of shame for eternity. I hope that you can find it in your hearts as decent human beings to accept my apology and as an humane act of contrition, allow us all to make a fresh start on this beautiful journey together. To you both I offer my hand as a gesture of humility and renewed friendship.'

Julie put her eyes up and Lord ATF shrugged, but both shook my hand. Awkwardness absolved, we trundled along to the hotel in happiness. As we neared Candolim I reflected upon my behaviour and how already India was driving me to distraction. It always does this to me. I lose my temper in an instant here and I would have to watch out. I would say that travelling in any developing nation is always frustrating to the point of serious temper loss on at least a weekly basis for most people. India does it to everybody at least twice as often and to me on an almost daily basis. It is the compendium of obstacles that conspire to prevent anything from going even roughly to plan that does it for me, add the heat, the smells, the noise and the general damn proximity to everyone else and you have a sure fire way to fire me up. At least a little part of me was hoping I would be able to conquer my frustrations by staying a while. Perhaps in time I could become blasé in the most trying of conditions, but for now I would settle for being just a little less highly strung. I would have to *make an effort*, there was no getting away from it.

Chapter Three

Goa, our new home. In many ways, India is the antithesis of everything I endorse, it is the exact opposite to everything I approve of, it is in fact my nemesis. For the benefit of anyone who has never set foot on Indian soil, I will describe India. The most uncomplicated way to do this is to simply say that it is exactly as it looks on the telly. It is dirty, smelly, noisy, sweaty, overcrowded and dangerous. The country is prone to natural disasters and to political and economical instability. It is a mysterious and superstitious nation, dominated by belief in brassy, multi-faceted deities. Everything moves too fast, too slow or too erratically. Everyone is too poor, too rich or too indifferent to notice.

It is fair to say that by and large, the general public in the West carry a broad perception that India is not a very nice place, and I would have to wholeheartedly concur with this. India is not a *nice* place at all, it is a fantastic place, an impossible place, an extraordinary place, an immoderate place and a most unlikely place for me to find myself.

As you have probably gathered, I am unequivocally an English Person, and as such I am hindered by my ramrod straight back and starched-stiff upper lip. Venturing into unfamiliar territory I tend to revert to type, becoming either the conceited Colonial Master or the thuggish Brit Abroad. It is fear that drives this behaviour, as it is with all my countrymen. It is a fact that the English are encumbered by their inability to accept and accede to different cultures, and I am no exception. India for me represents the ultimate in things foreign and it was ultimately this that drove me to visit for the first time, and partly facilitated my recent uprooting.

I see that it is time to share a few facts about myself.

I have mentioned that I consider myself to be a Mod. For the uninitiated, Mod was a 1960's fashion that grew from a bunch of working class London dandies with a proclivity for Italian style, American music and French newspapers. Posers to the last, the term is the abridged modernist, still the predominant intellectual movement of that era. I became a nine-year-old 'ticket' Mod in the 1980's revival that turned many Home Counties boys into wannabe Paul Wellers. And I've never really grown out of it, of course I conceal it as much as I can, but the others, they always know.

I was born into a working class Catholic family, the son of an Irish factory worker, an upholsterer. I grew up in a traditional Comprehensive manner with working class heroes and working class beer. However, unlike my childhood peers, I did not hate everything that was not working class, indeed, I had certain aspirations toward the middle and even upper classes. It was never material wealth I was drawn to, rather the sense of status, even *superiority*. I had always admired the gall of the toffs, their buffoonery and dumb belligerence somehow held an appeal for me, and I am not particularly proud to admit that it still does. There is something inherently funny about a genuinely posh person, something terrifically comical in the unswerving, brash self-confidence of the highborn. Hunter Thompson noted 'the genuine aristocrat never doubts his own worth' when referring to an upper class ex-pat Englishman in Rio whom he witnessed driving golf balls from the roof of his members club down into the slums below without a thought to the consequences. This image sums the whole nobility thing up for me - dangerous, disgusting, hilariously shocking behaviour.

It is with similar amusement that I view the Anglo-Indo Colonial period. I realise that it is acutely crass to make light of what amounts to nothing less than murdering invasion, theft and slavery, but (forgive me) I just see these puffy red-faced dunderheads clad in safari suits and pith helmets getting terribly cross when the *punkah* wallah isn't fanning fast enough or the char is late. It seems such a civilised fascism, in the vein of:

"I say, who's in charge around here? You sir? Right, my name is Curzon and I will be taking over from today. You will have to move in to the hunting lodge because I will be taking up residence in your palace. Ok, run along now my good man, oh, and *do* tuck your shirt in, the English are in charge now."

No doubt, further pursuit of Britain's pompous history was a contributory factor in my decision to visit India for the first time, but the primary factor was without doubt quite different. It was, in short, my religion, or to be more accurate, my aversion to it.

There are many clichés about being raised Catholic, the most obvious being the guilt we all feel for our Original Sin, and they are mostly borne out of true, common experience. However, there are other, more private, less well-documented experiences that come to us through Christianity's dogmatic education. As a child I had an abject fear of epiphany, specifically my own epiphany. There are many instances in Christian teaching where ordinary folk 'see God' and become devoted to Him thereafter. Indeed, growing up in an Irish community on the outskirts of London in the 1970's, there were no shortage of tales of folk from 'back home' who had had apparitions, seen moving statues of Mary, met Jesus while tending the turkeys in the field, etc. My fear was that I myself would see God and would have to be devoted to Him for the rest of my life. I was only partly afraid of the actual seeing God part, the real fear was the inexorable subsequent devotion. I figured that to dedicate my life to God would necessitate the Priesthood and a life of celibacy, no drinking in pubs and no fun. I wonder at how a seven year old can have been so affected and cynical.

My fear grew over time. I tried to hide by not looking at the statues in church but was then informed of God's omniscience and omnipotence and I knew there was to be no escape. I would see God in my dreams and would run away, disappointing God and inviting His Wrath. I considered being really bad so that God wouldn't appear to me, then I considered He would be more likely to want to *change* my ways rather than endorse them so I started to be really good to keep Him away. I began to fear breaking any of the commandments. I examined each one in turn in my Old Testament annexed Illustrated New Testament and felt comfortable I could keep them all; all except one - Thou Shalt Not Have Any Strange Gods Before Me. My anxiety was not based on me actually worshipping another God, this was most unlikely, rather I had told myself that to even *acknowledge* another deity was a terrible sin. It was in this way that I developed my innate fear of Hindu Gods, a fear that until recently remained with me. You see, they taught us about Hinduism at school, and of course they showed us pictures too. I had terrible dreams where I would be attacked by Vishnu or Krishna or Ganesh, and God (my God, Mr. Catholic White Beard) would be watching, shaking his head, tutting. As I grew up my epiphanical fears abated, but an irrational fear of Hindu Gods remained. In my teens, when I (ahem) smoked a fair bit of pot, I would frequently become paranoid and uncomfortable in conversations on the subject of India, or exposure to Indian music. I continued to have nightmares featuring a variety of gaudy, multi-limbed beings. Friends started to go 'travelling' in their gap years, and of course, India was always on the agenda. They returned with tails of stoned Saddhus, temples, whirling dervishes, *bhang lassis*. I resisted and wrestled with my dreams. One night I woke with a start and in a sweat, the dream had been too much and had become too frequent, I sat up in bed shaking and decided then that I must go to India to dispel these ghosts, a type of aversion therapy, confronting the fear head-on.

And so it was decided and off I went.

What I found differed somewhat from my expectation. Not only was I apprehensive about the whole 'God' thing, but I had travelled a fair bit throughout the rest of the world and was jaded with the 'traveller' or backpacker ideal of 'doing' various places and then harping on about them over a cheap warm beer in some 'exotic' location, flexing the old bullshit muscles. I knew the Indian tourist trail would be chock-full of tie-dyed wannabe spiritualists addressing anyone who would listen to how they had 'found themselves' in India.

In fact I was wrong, or at least partly, what I found was Goa, a package holiday destination much like any other. I found puffy red Brits, sleek tanned Italians and serious looking Germans with sausage brown fat bellies in tiny swimming trunks staring out to sea with their hands on their hips, just like everywhere else. And I found Catholics by the thousand, Indian Catholics. This I hadn't bargained for. But without doubt the most unexpected thing that I found, was in fact myself. Most unexpected. I shouldn't really go so far as to say that I 'found myself', that sounds too severe (not to say embarrassing), but I should say quite openly and honestly that I learned a great deal about myself in the few short weeks that I spent in India. I will also say that I experienced pretty much every emotion I had ever felt before along with a few new ones. India (I spent a week in Mumbai as well as Goa) was the most cathartic experience I had ever had. I learned through being exposed to this organised chaos, that I actually like dirty, smelly, noisy, sweaty, overcrowded and dangerous countries that are prone to natural disasters and to political and economical instability. I found out that I love mysterious and superstitious nations, dominated by belief in brassy, multi-faceted deities in which everything moves too fast, too slow or too erratically and where everyone is too poor, too rich or too indifferent to notice.

I found out that India is a prodigious expanse of confusion and that I love it. I found that India is a place that everyone should go to at least once, people say you either love it or you hate it, I disagree, I think you love it but you're too scared of what might happen if you admit it.

It is the combination of everything that I love about it. It is the funniest place in the world, no doubt about that. It is also the most frustrating, beautiful, filthy, serene, exasperating country I have ever visited. If you are considering going, stop considering and just go.

I once asked a friend who is extremely well travelled, which was the best country he had ever been to. His reply was immediate, 'India'. Why? I asked. 'I saw a bloke go to sleep standing up, what more do you need to know'. I think that about sums up why we chose to live in India. That and the fact that the people don't know what a Mod looks like, so they don't look at your Fred Perry, Loafers, Stay Press and white socks and say "Did you used to be a Mod?"

Oh, and it is sunny and dirt cheap, and we are tight. Indeed, The New Tight.

****

Here we are then, Candolim, Goa, India. For those who don't know, Goa is an Indian State, an old Portuguese enclave on the Arabian Sea coast. It is essentially four or five towns and a coastal belt of red-dirt, coconut palm villages, which now pretty much merge into one long strip of tourism (in the north at least). By Indian standards Goa is tiny; some 2000 square kilometres with a population of around 1.2 million. It is also very much un-Indian. Historically, Goa's mainstay industries were iron ore, fishing and agriculture. Today Goa relies almost completely on tourism. What was presumably a very serene coastline back in the heady days of the 1960's when hippies came to colonise the area, get stoned and dance around naked, ignoring the shocked locals, is now a full-on package tourist resort with a dangerously low water table and too many flushing toilets. I imagine that for many package tourists there is very little difference between Calangute and Magaluf, except that in the former there is next to no draught beer and the curries are better. This is what makes Goa so different to other parts of India - the multitudes of half naked farangs (foreigners) lolling about on beaches and careering around on scooters.

Tourism in Goa is not necessarily a bad thing. Of course it has serious ecological drawbacks, and it makes for a fickle economy that tumbles whenever politics renders the area less safe (Gulf War, Afghanistan, Indo-Pak relations, etc), but it does bring in money, something that India always needs. Poverty levels are comparatively low in Goa and comprise largely of migrant workers from other states, and literacy levels are very high, well above the national average. Goans are known for being laid back and easy going; they are predominantly Christian due to the Portuguese settlers, love sport, particularly football, and won't say no to the odd drink. Goa is an easy introduction to India. It is India Lite, India with *facilities*. Visit Karnataka, Rajasthan or West Bengal and you will experience India. Visit Goa and you will lay on the beach, drink beer, eat seafood, go to karaoke and get a taxi to a flea market once a week to bargain with locals dressed as Rajasthani peasants for tawdry gifts that are largely made in China. Goa is pastiche India with home comforts, and this is why we chose to station ourselves here. I have already mentioned that we are hardly intrepid pioneers; we felt that we needed to go to a place where we could revert to type whenever necessary, where we could be English from time to time.

The real reason for spending a year here was to create some *lebensraum*, some room to live. Julie and I had been working in London for twelve years and felt that we wanted a break, an escape, a rest. We had not chosen India because we wanted to discover ourselves or to discover India. We had no intention of *travelling* anywhere, we were coming to live. We also had no intention of experimenting with local culture, we were going to immerse ourselves in it in the way that we do in London, that is to politely say hello to the neighbours and close the door. We had come to a place that we thoroughly enjoy, where we are allowed to do largely as we please, where people are very friendly to us and where we can afford to live without working. That place is Goa.

Our intentions were similar: I was going to write the novel I had been attempting for the past five years and Julie was going to go back to basics with her artwork and spend hours drawing and painting for pleasure rather than deadlines. In addition, we would variously: take up Yoga, play darts, tinker about with old cars, play guitar, get drunk, sunbathe, learn a language, keep a garden, look for some way to help out in the local community, muck about, improve our photography, watch hours of cricket - that sort of thing. If you take out watching hours of cricket, and possibly the Yoga, these are things that can be done anywhere; they are in no way specific to India. The point is that these things are done *differently* in India. That's why we came, to do the same things we always do, but in a place where these things cannot possibly be the same, a place where even getting drunk is fundamentally *different*.

The only thing that I knew we could not do in India was paid work. There was no way that this would ever happen. The malevolent lure of the salary would not be dangled in front of me and the benevolent institution of work could not put its comforting arms around me. Twelve months without work, at least.

I settled into the lounger on the balcony terrace of the hotel room and contemplated this delicious fact. No work for a year. The next few weeks would be busy, trying to find a house, a car, a decent mosquito repellant, but ultimately it would not be 'work', it would not be toil, it would not be graft.

So we set about setting up home. We went in to see our friends who run the local bars and restaurants, put the word out that we were looking to stay for a while. First we found Anupam, the four-foot tall bartender of a four-foot tall bar, embarking on his eighth year of work in Goa as an itinerant Calcuttan. We had met him in previous years and were as delighted to see him peeking over the top of his bar as he was to see us.

"Hello my friends you came back you remember only me? Ha Ha." The ebb and flow of his singsong voice and appalling English was an elevating reception. I was also delighted to see that they had installed a large screen in the bar for watching football and cricket.

"Yes, we're staying for six months at least."

"Six months, oh my God, such a long time, so boring, what will you do?"

I hadn't bargained on the negative, "Not boring, it will be great, I am going to write a book."

"What book? What book writing is this?"

Neither had I prepared for this, "Err, it's err, oh it's just a story…never mind anyway, how are you? How have you been?"

Anupam slid back behind the bar to hide and reassume his demure coyness, "OK, I've been here only."

"What, you haven't been back to Calcutta after the season?"

"Yes, I have been there no?"

"What?"

"I have been there only."

"I thought you said you have been here only?"

"I have been there only, now I am here only."

"Right, I'm glad we've sorted that one out."

This is precisely the mode of conversation one can expect in India. Generally speaking, confusion reigns. I love this. There are a number of different types of interaction one can expect if using English as the communicating language:

- Speaking English with someone who speaks Good Indian English
- Speaking English with someone who speaks Bad Indian English
- Speaking English with someone who speaks Uninterested Indian English
- Speaking English with someone who speaks Absolute Nonsense

The difference here is in no way subtle, it is colossal.

Good Indian English is essentially perfect English. Perfect English that is peppered with sporadic archaic expressions, occasional comic book retorts and the odd example of irregular Germanic sentence structure. As most higher education in India is taught in English there are no shortage of people with whom to have perfectly normal conversations. These conversations only very occasionally venture into Battle, Beano or Beezer type exchanges, but do contain the occasional hint of an Albion Past. One only has to look at the English language newspapers for examples:

'Miscreants nabbed after burglary mishap'
'Dacoits raid swine farm, man dashed on fence'
'Gang of rowdies held after *paan* shop uprising'

Bad Indian English is primarily spoken with those who are either eager to please or eager to learn. Either way, you are their stooge. You will have to decipher and disseminate any information that may result from such exchanges yourself and you will also have to confirm any resulting actions or be prepared to bear the consequences yourself. It is not advisable to make any jokes or any small talk during such interactions, try to stick to basic communication, any whimsical meandering will certainly backfire. You will have these conversations frequently in India and will experience a varying level of anger and amusement at them, which will be largely dependent on the current temperature and the number of such conversations one has had in a given time period. Personally I like these interactions very much. They are a great test of patience and stamina. They are very often with waiters or just people who approach you in the street. For example:

"What would you like sir?"
"I'll have fish tikka and salad with one plain naan please."
"One fish is not there sir."
"Not one fish eh? (Chuckle) And the tikka?" (A terrible joke, a terrible mistake)
"Yes, ok sir, and for you Madam?"
"Err, I don't think I ordered anything, you said there is no fish."
"Fish is not there Sir but ok. With naan, yes, no problem."
"What is no problem? What have I ordered?"
"Sorry sir?"
"What have I ordered? What have you written on your pad?"
"Pad sir?"
…and on and on it goes. Or perhaps you may be approached while

waiting for a bus, a train, a vacant toilet, anything really:

"Hallo, what is your good name sir?"

"Oh hello, my name is Mick."

"Mike, very good Sir. What is your purpose here?"

"Very difficult to say, philosophers have been trying to work that one out for years. Do you mean in India? Or here at this bus stop?"

"Very much Sir. You are philosophy no yes?"

"Err…"

"Please come to my house sir, I will be honoured to serve you tea and biscuits sir."

Uninterested Indian English is fundamentally terrible English with which Indians who don't really speak English will repetitively attempt to sell goods to foreigners who often do not speak English themselves. It is usually a one-way conversation comprising terms such as:

'Only look.'

'Cheap price.'

'Something?'

'Hallo orange?'

'Look at my shop.'

'Boat trip?'

'Remember me?'

'Hallo Sir, drums?'

'What you want?'

'Taxi?'

Any attempt to elaborate into areas that do not involve the sale of any goods or services will most often result in repetition with the occasional hopeful variation from the vendor. An example might be:

"Hallo Sir, look at my shop. Very cheap price. Only look. Looking is free."

"I don't want anything thank you, but do you know where the post office is?"

"Very cheap Sir, only look, only look. Come"

"No thank you, do you know where the nearest post office is? I have a letter to post" (showing the letter).

"Sir, look at my shop Sir. Only look, looking is free. Very cheap Sir, look."

But speaking Absolute Nonsense is by far the best. It has no recognisable value for either party save for joy, mirth and an overarching feeling that the world is fundamentally a good place. It is sometimes simply miscommunication, which can be variously both enjoyable and irritating, but more often is just plain weird and is at its best when the general gist of conversation can be followed but the purpose behind the interaction remains wholly unfathomable. A shining example came from a poolside waiter, a tall, scruffy, forty year old man from Ooty called Felecious, or at least this is how he said his name. He had the most broken shoes I have ever seen worn and a remarkable take on English. I really enjoyed the man's directness, and went back to the same establishment many times to experience it further. A recent conversation between us went:

"I go to Jew, you go to Jew?"

"I'm sorry?"

"You go to Jew, animals are there, many animals are there."

"Oh, the zoo? No I have not been."

"One big monkey is there, like me. Children's food is bringing to cage, hand comes out, hand comes out for food." (Thrusting his arm out in front, palm up and open, looking crazed with wild bulging eyes, serious as anything).

"You mean a gorilla?"

(Phonetically) "Goo-rill-er. Yes, maybe. Hand comes out for children's foods."

"I expect its hand comes out for the food." (What am I saying? This is surely just what he has said, I'm falling right into his trap)

"No, hand comes out, hand comes out, HAND COMES OUT. (Thrusting his arm back and forward at an insane pace) NOT food. Hand comes out."

"What? Err, sorry?"

"HAND COMES OUT (shouting now, screaming even, so that other people are looking round, frowning) HAND COMES OUT. You want cup of tea?"

"Err…yes please. Thank you."

To be fair and to veer back from semi-jingoistic piss taking, India is one of the easiest and safest countries in which to travel or reside if one speaks no other language than English. In any given situation the English speaker can expect to be no more than about five minutes away from help. Go into any Indian city and ask for directions, very quickly someone will be found who speaks English and is very willing to help. I am not suggesting that they will be able to help, merely that there will be no shortage of people *willing* to help. And these people are unbelievably honest, most of them are not even on the make, despite the clear wealth gap. In a country containing one sixth of the world's population and a lion's share of the world's poverty there is next to no crime. Property break-ins and robberies make national headline news. On page three of the Times of India in September 2004 a headline ran 'Another Mobile Phone Theft' above an article referring to the third mobile phone theft in as many weeks in the city of Pune. Four million people live in Pune, at least half a million have mobile phones. Just think about that for a minute.

****

Our Candolim barman, Anupam, is a small twenty eight year old person of honour. He has worked in the tourist industry in Goa since he was twenty. He earns a basic salary of around 100 rupees per day and relies heavily on his tips. He is responsible for running a sports bar and he takes his job seriously. In the main, his customers are Europeans who drink beer and watch football, so of course, they are, in the main, British. Booze, football and Brits bring agro and Anupam gets his fair share of it. For a man not an inch over four feet he handles himself and the situation very well. He is in short, (and he is short) a thoroughly *decent* chap. He treats all his customers well, attends to their needs, shares a great knowledge of world football, cricket, rugby, in fact pretty much all sports with them and makes them feel at home. He wears a 1980's era West Bromwich Albion shirt most days and sits patiently behind a bar that he is unable to see over, waiting for whoever may come in to get drunk, make fun of his English and eventually leave. That is an average day for Anupam. His tolerance and humility is indicative of the Indian worker, often performing a thankless task for very little return, without complaint. Anupam's gentlemanliness is in no small way responsible for us choosing Candolim as our station. After all, a sports bar is a sports bar and there are not too many of them in India, even less with a staff of such fine reserve.

As I have said, Candolim is largely a package tourist area. It is prime beach-belt, sunshine strip; packaged paradise. Sandwiched between the busier, scruffier Calangute and the distinctly upmarket Fort Aguada, it is chiefly avoided by the younger crowd and frequented by the older, fatter, slower tourist. It hosts a plethora of safe, dull restaurants and bars serving bland multi-cuisine dishes accompanied by Elton John, Cliff Richard and Brian Adams. Candolim has cashpoints, security guards and air-conditioned hotels. According to the locals it is *quiet*. According to the guidebooks it is *popular*. Accordingly it is somewhere in between, and for us it would do. There is not that much to do, and that would do nicely.

We began to settle ourselves. We lounged at the Casablanca Hotel for a few more days while acclimatising. We dined at the myriad lacklustre restaurants and swam in the sea. We worked on our suntans and drank beer. After a couple of weeks of meeting people and explaining our intentions we set about finding a home. There are not really any estate agents in Goa, there is only 'asking around'. Asking around is frustrating and time consuming. The situation reminded me of starting out at college - the first few weeks is spent making friends and the rest of the time is spent trying to get rid of them. We flirted with the British ex-pat community, an established group of forty to sixty-somethings who own and run bars, restaurants and hotels in the area. We found absolutely nothing in common with them. Not a thing. We found nothing wrong with them, we just found their version of home from home very, very dull. They charged slightly more than the locals and consequently catered only to their own countrymen. Of course they employed the locals and brought money into the area, but they seemed strangely disinterested in anything other than their own little world. We asked them about where to live and they offered only where not to live. They sang karaoke, moaned, drank a lot and that was about it.

We started to frequent places run by locals to ask about accommodation. The response was always the same - "I'll see what I can do", nothing materialised. The problem seemed to be that we were looking for a house and not a flat or apartment. We wanted somewhere to *live* and not just somewhere to stay. It seemed that unlike the Englishman, the Indian's home is not his castle. No one could understand why we wanted more than one bedroom. Eventually we were put in touch with a young man called Jude who pertained to be an estate agent. He took us on the back of his motorbike to see numerous places but none were quite right. We were hoping for a beachfront property but quickly realised none were available. He took us to see a fantastic old Portuguese house with a walled garden full of fruit trees and two rather splendid plaster policemen standing guard at the gate. It had been vacant for some twelve months and was filthy but we were ready to take it on the basis that it would be cleaned thoroughly for us. It seemed perfect: large, authentically Indian with a European feel, secluded, safe even. But I blew it, I said to Julie, "There's just one thing, it reminds me somehow of The Shining." That was the end of that.

Dejected and frustrated we turned closer to home and appealed to our hotel owner friend Gurmeet to help us in our plight. He took on the task with aplomb, taking us to numerous places in his tiny white Maruti van with 'Spice Girls' on the side, but again to no avail. They were always too small, too dark, too dirty or too weird.

"Gurmeet, there must be a place for us here, we can't just turn around and go home, there has to be somewhere."

"Well there is one place I've heard about but it's not finished yet, do you want to have a look?"

And there it was. As we rounded the corner I saw it and knew. A brand new orange bungalow standing slightly raised up in its own grounds, set back slightly from the road. It was a mess of angles, a cubist chaos, making a mockery of modernism. Who the hell had built this? We met the owner, architect and builder, a slight, grey man with a stern, worried look and a strange faded tattoo on the inside of his lower arm. We negotiated a price and took it there and then. Two bedrooms, two bathrooms, kitchen, living room, useless middle space and a roof terrace over the entire building with a large marble staircase leading up to it. It would be ready in two weeks. Our new landlord, Mr. Arnold Fernandes, had apparently been building it for the last two years on a plot owned by his wife's mother. Time and money had got the better of his ideals and he was in considerable debt over the project. As we shook hands on the deal after some stalwart bargaining on his part, which entailed him not budging one inch on his somewhat inflated price, he broke down in tears. The relief for him was too much and he was unable to speak. He finally managed some words , "You can trust me…" he blurted before the tears returned. We went to his house for a glass of port wine and to meet his wife and family and do lots more hand shaking.

"You will come to the inauguration next week?" he asked.

"Yes of course" I replied, not having a clue what this entailed. I didn't care, we had found our place.

We returned to the hotel to celebrate our find and get on with not working. We had two more weeks in which to not work before we would move in to our new home and not work some more. At the hotel, the water was turned off and the electricity wasn't working. Nothing was working, least of all us. We raised glasses in a toast.

The next morning I went in to an Internet café to check mails on things back home. I opened an email from an old work mate that I should have ignored.

"Hello Mick, how are you doing? I hope you're not working too hard, Ha Ha."

(He also used an exclamation mark or two, but in an effort to offset the over use this punctilious little shit, I refuse to use one, ever.)

"Actually I've got a little job for you. There is a call centre in Bombay, a company called TDS that is starting to do market research and they want someone with experience to go and oversee the project. Nice little consultancy role for you. The boss is called Avinash, I met him here in London a couple of weeks ago and I told him you were in Goa. He wants to meet you, you should do it, sounds like good fun…"

And there it was, the thing that could never happen had just happened. As I closed the mail I knew already that the job was mine and that I would take it. I find it hard to describe how it felt. I imagine it is how a failed addict feels when they take the first fix after a lengthy abstinence, a terrifying blend of euphoria and self-loathing. An elated sense of dread; a healing sickness.

And so it was that in less than one month off the Work, I was thinking of getting right back on it again. I had taken an acquiescent genuflection at the altar of inevitability. Oh kindred mortals, how powerless we are.

Chapter Four

Arnold Fernandes is the small, smiling head of a small smiling Catholic family. Unlike the majority of the English and Irish Catholics that I had been bought up with, Goan Catholics practice their religion fastidiously and display their beliefs ostentatiously. Arnold had made a major point of incorporating his religion into the structure of his house by including welded crucifixes into the window grilles and fixing alternate cross shaped stone banisters on the external stairs. He had also built a petite glass shelf in the living room with a single power point below it to house a cruciform night-light and several statues of Jesus and Mary. I had offered myself as a Catholic on the basis that he might be more willing to let the house to us - this was a mistake. His assumption had understandably been that I was a *practicing* Catholic in the Indian mould, not just another Graham Greene character battling with his guilt from sporadic church attendance. It was on his insistence and our trepidation that we eagerly agreed to leave all religious regalia intact upon occupation. It seemed our new house was also to be the House of God.

Having built his house slowly, Arnold had finished it quickly with the sudden injection of funds that our down payments provided. Before we were able to take residence, it was agreed that he and his family would spend two nights in the property to enjoy their creation briefly before handing it over to us, but no-one would be moving in before the inauguration ceremony, no sir, not a chance. This would clearly be inauspicious in the extreme and probably impious too. I had guessed that the inauguration would more or less be a party with a priest present, and I was fairly near the mark with my assumptions. I had however, not bargained on the length, formality and sincerity of the piece.

Indians seem to love to attend openings. Browsing any local newspaper will throw up reports of numerous inauguration parties for openings of new businesses, homes, swimming pools, garages, football pitches, anything really. It will also reveal any number of small adverts taken out by well-wishers in support of these *debutees* and *debutantes* and their new premises or ventures. I once asked a taxi driver what a crowd gathering in the middle of a roundabout were waiting for and it transpired that the 100 strong congregation were in place to celebrate the turning on of a new electricity sub station generator (a delicious irony prevailed as the town experienced a perfectly coinciding power cut which delayed proceedings for some two hours). The earnestness with which these occasions are imbued is stultifying for the casual observer, like watching paint dry. Arnold's home inauguration was no exception, and at one point, that is exactly what we did, we watched paint dry.

The house is a bizarre take on modernism, a sort of Gropius Vs Gaudi fight, where both die in the skirmish and the Clerk of Works for the local council finishes the job off with a sense of public duty and uninvited *tradition*. A flat, square, protuberant base sitting heavily on the ground, flanked by two jutting galleries to the front elevation with an external grey marble staircase to the right, leading from the veranda to the upper tier roof terrazza where one might lay in the unforgiving, relentless sun and burn. There is little restriction from building regulators in India, and provided the correct palms are greased, there exists a real opportunity to experiment with the private house. Arnold did his own drawings and had them approved and corrected by an architect. He then supervised the building of the house with little intervention from anyone 'official'. And he did a remarkable job. Surrounded by a mix of red soil and building sand, and set up high on its little hillock, the house looks like something from the Sierra Nevada: an orange ochre monolith with a large white concrete-sculpted Christian cross out front to remind strangers who is boss in this town. A desert residence almost? A *Des Res*? Certainly. The house was to be officially named "Vailankanni" which is the name of an Indian town, site of a famous apparition of Our Lady, and unofficially known as "Isabel's Nook" (in fidelity to Arnold's wife Isabel, for whom he had built this monument), and both appellations had been skilfully painted on the fascia above the front door.

As we arrived for the party, Arnold and his son Jack were standing on the veranda steps watching a sari-clad servant furiously sweeping the 'garden' while a sweaty man in a stained vest feverishly whitewashed the concrete cross which steadfastly denoted the house as Catholic among its Hindu neighbours (the Hindu equivalent to this is very often a rainbow coloured turtle with a planted tree on its back - so much more fun). Arnold smiled his apprehensive toothy grin.

"Hallo, come, come, come in," he bid, waving frantically toward the house.

"I see you have had a large cross made", I ventured, nodding toward the concrete structure, smiling weakly.

"You don't like it? It must be there, we have to have it." He was resolute in offence.

"No, I like it very much, I am very pleased to have it there," I broadened my smile to bolster the lie.

He softened, "Good. You are early, come inside." He turned to enter the house and we followed. I resisted my inclination to contest this wrongness despite my certainty that we were not early. We had been told to come at 3pm and it was already 3.30pm, but I had relinquished disputes over timing by this stage for my own sanity. I am never late or indeed early without explanation or apology under normal circumstances, but here in India this appears to be abnormal, so 'when in Rome' and all that…

We entered the house to find it cleaned of every spec, and potential spec, of dirt or dust, a real accomplishment given the omnipresence of Indian dust, the daily descent of which swathes everything. The carroty floor tiles gleamed in the sunlight. The chocolate brown gloss of the window frames bulged and shone, sticky in the heat, and the surprisingly quiet, spanking new ceiling fans caused the numerous curtains to billow and flap against the soft yellow walls in the semi-silence. Inside we met Isabel and her daughter Catrina, or Sweetie as she was playfully christened. Isabel hugged us both; an undersized tub of a woman with a bowling gait, she was smiling for India today, and her grip was sincere. Sweetie offered her limp hand to shake and backed off coyly, grinning at the floor.

"You're early, have a cup of tea." Isabel turned back to the kitchen.

"I thought it started at 3? It's already 3.30." I couldn't help it.

"But it is early in the day no? People will not come until later, until the priest arrives. Have a cup of tea."

Oh God, the worst possible outcome - being very early. I have no taste for formal parties at all and I cringed at the thought of an agonizing wait for people to arrive and the innumerable introductions and resulting small talk that would permeate the whole afternoon. Add to this the novelty of being probably the only white people at the party, and the attention that this brings, and it becomes the most tedious exercise imaginable. I had hoped for a quick blessing from the Priest, lots of smiles, hand shaking and sympathetic nodding and a sharp exit on some pretext or other. I needed to know the order of things so that I could formulate Plan B.

"The house looks beautiful, you must have been cleaning all week. What time does the Priest arrive?"

"He should be here soon, Arnold will pick him up from the church."

"How soon, maybe one hour or more than that?"

"Yes, probably."

Isabel handed me a cup of tea as I tried to think how to pin her down without being too obvious. Plan B was beginning to take shape when...

"It is Mummy's birthday today." Sweetie chimed in unexpectedly.

"Oh, we didn't know, congratulations." Julie moved forward to kiss Isabel, leaving me in the routinely awkward position of not knowing whether to kiss someone or not. I'm hopeless at this, habitually bungling things by presenting a hand for shaking into the midriff of someone who is looming for a hug, or worse, attempting to kiss someone on the cheek who is visibly withdrawing with an outstretched hand and a rather stiff countenance. I never decipher a person's intent and in the absence of a prescribed protocol I am always the fool. In foreign countries my indecision is somewhat inflated, and in India, with all its prudish sexual inconsistencies I am likely to be made a royal buffoon. I outstretched my hand quickly and bowed as I shook. Success. Success on the non-embarrassment front at least, no chance of a Plan B now though, we couldn't duck out of her *birthday*.

We took our tea back out to the veranda to speak to Arnold again.

"You didn't mention it was Isabel's birthday, we would have bought gifts. Did you purposefully arrange the inauguration to coincide?" I asked him.

"Yes."

Conversation with Arnold can be like pulling teeth at times.

People started to arrive and the greetings began. Brothers and sisters, aunties and uncles, cousins, father cousins, mother cousins, brother cousins, sister cousins, neighbour cousins, neighbour uncles, neighbour aunties, other cousins. And then the children; many, many children. Having met everyone at least once I felt exhausted to be so in demand, everyone wanted to talk to us. Our popularity stemmed from one salient irregularity in the banality of the discourse – everyone wanted to know how much we were paying to rent the house. Behind a thin veil of genteel dialogue a burning absence lurked, they *needed to know* how much Old Arnold was making out of the foreigners. Interactions typically went something like:

"So, you'll be moving in to this house then, how long will you be staying in Goa?"

"Oh, between six and twelve months, we haven't decided yet."

"Its such a nice house, so beautifully finished. I'm sure you'll be very happy here. HOW MUCH IS HE CHARGING YOU FOR IT?"

Very quickly we developed a lie. We figured if we told the truth (15,000 rupees per month, around 200 pounds, a fair sum we felt but well over the odds as far as the locals were concerned) not only would they think we were mugs, but would probably also try to tap Arnold for any debts he might have with them, the likelihood of which was high.

"Well, I'd rather not say because it is really between Arnold and ourselves, but it is in the region of 8,000 rupees."

"Oh, I see, well, enjoy your stay." And they'd be off, disappointed and in the main disbelieving.

The priest arrived, squashed into the passenger seat of Arnold's Green Maruti 800. As he stepped out of the car, expanding into the available space like a sponge, he loomed tall. The cleric brushed down his vestments and viewed the awaiting congregation with an aloof smirk. Arnold's son Jack lit a bundle of firecrackers on the foot of the newly whitewashed concrete cross and the crowd observed the pyrotechnical display with indifference and without joy (as far as I'm concerned Indian fireworks are always a joyless affair, all noise and no display, particularly in daytime, and they're just so frequent and recurrent, so damn ubiquitous, I have seen people lighting fireworks alone in broad daylight, wincing at the noise and walking away without so much as a smile, where's the fun in that?). The priest looked almost perturbed by the uncouth display.

We followed the clergyman into the house in sombre procession. Possibly one hundred people attempted to occupy the living room and I was reminded of an increasing yearning for personal space. We had been in India only four weeks and the proximity of other people was already beginning to rancour. I consoled myself that in a few days Julie and I would be the only people in this space as we settled in to a cramped corner to endure the priest's blessing. And how he blessed. He blessed and he blessed. The duration of a Catholic Mass is around thirty minutes if the homily is omitted, this inauguration ceremony lasted double that. The priest seemed to summon the support of the entire inventory of saints and then cover the walls and the congregation in scented water, hoisted from an aspersorium while holding a prayer book aloft and chanting, eliciting a response in Konkani which was most likely 'pray for us'. The water splashed on to the newly painted walls and appeared to be staining rather badly. I scratched the paint with a finger and watched the powder appear under my nail - cheap Indian paint, the scourge that renders a whole country tatty. We stayed in our corner politely pretending to pay attention to the solemn event, but were clearly wavering after around twenty minutes. The priest finally relented in his anointments and commenced a non-baptismal prayer. Remaining superficially alert, I began to ponder the wetness of the walls - would the water leave a stain? Would our pristine living room look like a drenched Jack Pollack canvas? The fans and the temperature began to have an effect and I shared the opportunity with Julie - we watched with considerable interest as the wet paint slowly dried, enjoying the drab spectacle, hoping God wouldn't be too angry.

After around forty minutes the rest of the gathering also began to waver in concentration. I noted with envy those who were positioned near the door who were sneaking out to the veranda, reaching for cigarettes. It seems the differences I had observed between English and Indian Catholicism were not as acute as I had imagined.

Finally the priest relented and proceeded to bless the other rooms, the crowd parted but only the immediate family followed. We escaped to the intense heat of the roof terrace for a frown of respite. After shaking everyone's hand again and thanking the family and the priest for such an enjoyable spectacle, we made our escape. Arnold dutifully escorted us to the road and as we parted he declared:

"The food will arrive at around seven, be sure to be back by then."

I tried to think on my feet. "Oh, the party is going on until tonight is it? That's a shame, we have arranged to meet some friends for dinner, we won't be able to make it, I'm sorry, I'm sure we'd have really enjoyed it."

"Surely you can make an exception tonight, this is a very special day for us and we would be unhappy if you were unable to make it, you are going to be our first house guests, it is important that you come."

"Err, I guess we could rearrange," I looked to Julie imploringly but she turned away with cowardice, "err, we'll be back around seven then."

"Good." Arnold turned and walked back to the house. We walked on in silence.

As we approached the house again, the sun was beginning to set and the mosquitos were beginning to rise. Turning the corner from the main road we heard the music for the first time. Ear blisteringly loud music.

Noise is part and parcel of Indian life. The best piece of advice I could give anyone is to bring earplugs. Pretty much everything is noisy - car horns, worship, dogs, even the cows make loud belching groans. But music is exceptionally loud. It is *interfering and overloading* loud. It is turned-up-to-eleven, completely-distorted-in-the-speakers loud. Goa has a curfew on music, bought about by a recent pandering to the moneyed package tourist and an attempt to curb the non-spending backpacker set and their un-taxable partying (a futile gesture based on nonsense, the former are even more shrewd, wily and skint than the latter, they come to Goa and don't spend a damn penny). Unless you are a policeman or know a policeman very well, you will have to curtail your party after midnight. Before midnight, however, one is at liberty to PLAY MUSIC AS LOUD AS ONE LIKES, and people recurrently do. I am unsure as to exactly why people play music so loud as it is clearly not for enjoyment. Perhaps it is simply because they cannot do so after the curfew and feel obliged to do so before it; but they always do it nevertheless. Arnold's party was to be no exception to this unwritten rule.

Half way along the approach road we recognised the song among the distorted sound. We were being treated to music from the Lion King. We were already apprehensive about this party, but Jesus Christ, the fucking Lion King? We could only hope that it was for the young children and that things would change.

Having reintroduced ourselves to everyone who was still there, all of whom were still sober, uninterested and uninteresting, and having once again confirmed for them all that 'Yes, we were only paying around 8,000 rupees for the place', and that 'No, Arnold wasn't paying any extra to our friend Gurmeet who'd introduced us', we settled in to listen to the rest of the soundtrack to the Lion King at some considerable volume. After some very shouted conversation with Isabel's sister who was visiting from Canada, which went broadly on the lines of, "What?", "Sorry, you'll have to speak up, the music's a bit loud", "Yes, we are from England", and "No, never been to Canada", "No, not first time in India", it became too much and we made a dash for the roof to escape the noise. The roof terrace had been turned into some type of auditorium for the evening, rows of plastic chairs facing in the same direction. Around thirty people of various ages sat in disparate groups facing the front and generally not talking or doing anything. After around ten minutes on the perimeter of this non-audience Julie and I still could not fathom what was going on. Beaten by the arcane inexplicability of the scene, I finally summoned one of a group of five teenage boys who were sat together to ask what was going to happen and why everyone was sat in rows, he looked at me nonplussed and said 'Nothing will happen, party is here', and then returned to his friends who giggled amongst themselves as we went back downstairs confused and perturbed.

Noticing that the speakers were pointing out of the windows, I went into the house where the music was unsurprisingly quieter, and found that most of the young people were in the back bedroom where the music was at an almost acceptable level. It seems they were playing music so loud outside in order that they could hear it sufficiently inside. There were about twenty youths aged between fourteen and twenty, who were all sat on the bed singing along to the Lion King. I turned away in disgust and collared Arnold.

"Do they really like this music? They seem a little old for this sort of thing."

"Lion King is everyone's favourite, you don't like?"

"Err, no actually, not my favourite if you know what I mean. But, you know, fair enough, whatever people like I suppose." I dashed back outside before he could ask me what I would like to hear from what I knew would be an impossible selection.

Back outside the music finally finished and I attempted to strike up a conversation with Arnold's cousin about his mobile phone. In the brief moments before the Lion King was again presented vociferously, I asked him about the screen picture on his phone which appeared to be of a sexy woman.

"Who's that on your phone then eh? Eh?" Doing the 'nudge nudge' bit.

"It's Mother Mary." He replied showing me a digital representation of Our Lady, Jesus' Mother.

I won't be asking that again.

Food was served and the ear bashing relented momentarily. Isabel is a fantastic cook and she didn't let us down. *Chana Masala*, Chilly Chicken, *Biryiani*, Pork Vindaloo. Mountains of it. In this situation it is impolite to have less than two plates full and we readily obliged. When we could really eat no more, we reclined on the veranda to digest. The thunderous, roaring Lion King recommenced. After a shared beer and a glass of port wine it was time for the Happy Birthday ceremony. The assembly was arranged in the living room and The Lion King was paused. A cake was produced and Happy Birthday was sung to a contented looking Isabel. In India, Happy Birthday has numerous verses, sung to the same tune, this rendition included 'Happy Birthday to you', 'May you have many more' and 'May God bless you.' After the singing, cameras were produced and a line formed by family and close friends who each in turn, fed pieces of cake to Isabel, pausing for photographs of each bite. Mercifully the proceedings halted after about ten people and everyone was spared the spectacle of our discomfited inclusion. The Lion King once again prevailed.

Eventually people started to leave and we took our opportunity.

"Thank you so much for such a great party, the Lion King has really started to grow on me", I said to Arnold and Isabel together, "I can't wait to move in. When should we arrange to move in?"

"How about Tuesday? It is Sunday today, we'll stay for two days and you can come on Tuesday."

"Great, see you then."

As we departed relieved, they waved us off accompanied by the diabolical chimes of the first few bars of the Lion King.

****

Before moving in we had to do some shopping. We had agreed to take the house only partly furnished as it was clear that Arnold did not have enough cash to fully kit the place out. It would be up to us to buy a fridge, cooker, TV, and all the other household items necessary for a comfortable life. Arnold insisted he would help us acquire the said items, although why we would want his help I was unsure. Buying white goods, or 'wife goods' as Julie likes to caustically entitle them, should be a pleasurable experience, but it never is. High price ticket items are rarely easy to buy; consider the hassle of buying a house, a holiday or a second hand car. In the UK, fridges and washing machines are bought from out of town warehouses from bored, indolent staff who have no information about and no interest in their wares. In India, or in Goa at least, things are somewhat different.

Arnold picked us up in his car and drove us to Mapusa where he knew a man with a shop. He muttered something about previously working at the shop but after four or five further questions it was still unclear who worked for whom and I decided to leave it, maybe it would become obvious when we met the shop owner. I had tried to resist getting Arnold involved because I knew he would try to help but just end up getting on my nerves, trying to do everything for us, I told him as much but he insisted, "No, I will help, you are staying in my house, you are my responsibility, you belong to me now." What can one reply to such conviction? I just left it and hoped the trip didn't turn in to our first showdown. We pulled into a small shopping complex and parked next to a sign saying 'No Parking for Cars'. I pointed out to Arnold that there were numerous empty parking bays immediately in front of the shops and pointed to the sign, he dismissed my ruminations saying, "This is better, I know the man." Again, I just left it, what did I know? We entered the small shop, which was packed with both electrical goods and electrical goods salesmen. We were introduced to and shook hands with the proprietor who seemed genuinely not to recognise Arnold as either a friend, previous employee or even an acquaintance. Arnold spoke to him in Konkani and he smiled at Arnold and then at us. I found this most unnerving - it happens so often in a foreign country that I should be used to it by now, but my suspicions are always raised and I become sure that a rip-off is imminent. I stepped in to take control.

"We need both a fridge and a TV, can you show us a few of each please?" I stepped in front of Arnold to remain in charge, turning my back to him each time he tried to hop back into the action. He was visibly disappointed and I had to remind him that it was our money we were spending and that we were perfectly capable of choosing items ourselves. The salesmen turned out to be even worse than Arnold, clamouring to show us everything in the shop at once. I have never seen such excitement in a shop. Radios, toasters, food mixers, battery chargers, kettles; coming at us from all angles. At one stage Julie lost control, letting out a shriek of laughter as a slender, serious man with greasy hair loomed round the corner presenting an upright washing machine, just his head visible, "Washer sir? Madam?"

"Look, stop all this nonsense, one at a time please, the TV first. Can you show me a number of TVs, I'm most interested to hear the sound quality, picture is less important. Show us the three cheapest and we'll pick one. I'm not interested in aesthetics." I turned towards the TV man.

He did not understand me and looked to another salesman. "Yes, sir?"

I repeated my statement more or less, but simplifying it considerably. Clearly he did not understand but proceeded to show me the most expensive TV in the shop. I resorted to basics, "Tell me how much each TV is." I said, pointing at the ten or so TVs on show. Once we had ascertained which were the cheapest I asked for demonstrations, "Ah, not possible sir, electricity is not there."

I turned to Julie. "Can you believe the irony? We've come shopping for electrical goods and there is a power cut. We'll have to come back, oh for God's sake…"

"Let's just pick a few things out and come back later when the power is back on."

Julie is most sensible at times, and unlike myself, quite able to cope with the odd little setback without blaming the World. But the Indian electricity supply must *be* the worst in the world. Power cuts are regular, if not daily, and in a rural area like Goa often last for hours. I have been to numerous 'developing' countries, many far less developed than India and have never encountered such an unreliable power supply. I have spent the last five years or more trying to uncover the reasons behind such an unstable grid and no one seems to know why it is so bad. I regularly ask people whenever there is a power cut "Do you know why there are so many power cuts? Is it that the system is so old? Or that it is so disparate? What is the history behind it? Who installed it? Why can't it just be reinstalled to a higher standard? What are the issues here?" But nobody seems to know, the only reply I get is something vague about load-shedding, that 40% of electricity is stolen and corruption is to blame. Sounds like the problem is here to stay.

We selected some likely TVs, decided on a fridge, toaster, voltage stabilizers (to protect the aforementioned items from common power surges) and some floor standing fans which we hoped would contribute to airflow given the lack of air conditioning in the house and the desperately sticky post-monsoon humidity. With this done we set off round the corner to locate some mosquito netting to be fitted over the windows. I love the fact that one is not expected to install anything oneself in India, everything comes with a tradesman at little or no extra cost. I gave the shop owner measurements for the windows and he quoted me a price that included stitching Velcro to the nets, delivery and someone to attach Velcro to the frames. The price came to around ten pounds for the whole house. Do It Yourself will never take off in India.

Next stop cutlery, crockery, pots, pans, bedding, towels and all the other necessary life tackle, all in one shop, and less than thirty pounds. Kitting out our home was becoming quite enjoyable. We returned to the electrical store and filled out all the necessary warranty forms for the goods, chose the TV, paid and left. On a whim, we also bought an electric oven that we were assured would run from the standard electric socket. We returned home happy and took delivery of the goods within a couple of hours.

The TV was appalling, terrible sound quality, one of the fans did not work and the oven needed a higher electricity output. No problem, one phone call and replacements were bought for the fan and a different TV that was better than the first. The oven however presented a problem. It seemed that items could not be returned if they were not faulty, so a return to the shop in person was necessitated. I took the oven in an auto rickshaw and entered the shop.

"Here is the oven, as I told you on the phone it will not run from our power supply so I would like my money back please."

"What is wrong with it sir?"

"Nothing is *wrong* with it, but it is useless to me, I cannot plug it in."

"This is not our problem sir, if it is not broken we are unable to replace it. All ovens are the same."

"I do not want it replaced, I want to return it and get my money back. What do you expect me to do with it? Re-wire the whole house to accommodate it?"

The assistant cowered at my raised voice and scuttled into a back room. I approached the manager and repeated my request.

"See sir, we cannot give your money back if it is not faulty. You will have to keep it."

"For God's sake man, do you think I am stupid? It is of no use to me, I have not used it, I cannot use it. Just put it back on the shelf and re-sell it. I demand my money back." I was by now demonstrating visible rage.

The manager ignored me and took a phone call. Another assistant attempted to intervene and I showed him the palm of my hand with some vehemence, he retreated and stood behind the manager. The manager finished his call.

"Well, what are you going to do about this? Don't underestimate me, I will not leave this shop without getting my money back." He looked at me tiredly, wanly. I squared up to him, "Look, I am not moving from here until I get a refund, there is your damn oven, now where is my money?"

He returned to his desk and sunk his head in his hands. "This is not the way things work here."

"I bought a large number of items from you yesterday, I am what is known as a Valued Customer, does this count for nothing, why are you treating me this way? Do you want me to tell the other customers in the shop? Do you want me to start harassing the other customers? I will become your worst nightmare if you do not give me the money immediately."

"I can give you a note for credit."

"Absolutely not. Give me the money, NOW." I was really shouting now, slamming my fist on his desk, no doubt veins bulging in my neck.

I sat on a chair to calm down as the manager slowly brought a chequebook out from a locked draw. He wrote me a cheque for cash, it took him at least ten minutes to write it, he might have been writing it in blood. When it was done I snatched up the cheque and stormed out without another word.

Back at the house Julie asked me how it had gone.

"Err, lets just say if anything else goes wrong with the stuff we've bought from him, you'll have to deal with it."

The next day I turned the TV on and a loud popping sound emitted, followed quickly by the picture disappearing into a small white dot in the middle of the screen. "Err...Julie?"

Julie called the shop and they sent a repair man immediately. No arguments, no grudges, no animosity. Once again I felt embarrassed by my bullishness. I considered my temper and made a note to myself to return to the shop to apologise next time I was in town. The repair man arrived on a motorcycle with a small bag containing a couple of screwdrivers, wearing flip-flops. He detached the back of the TV and removed a large circuit board. He asked me for a newspaper in which he wrapped the circuit board, positioned it under his arm and got back on his motorbike, saying "tomorrow" as he rode off. Julie and I looked at each other.

"Are you sure you explained properly on the phone? An unknown, unidentified man just rode off with half of our TV under his arm, I don't fancy sorting this one out if he doesn't return, what if he drops it? What if he never comes back?"

"Oh, don't be so melodramatic, he'll be back tomorrow like he says."

Next day he arrived as stated, refitted the part, fixed the TV and left without a word. Things work differently here, I had better get used to it.

We settled in to the house and set about meeting the neighbours. This was not difficult. They came to us. They had to; they *needed to know* how much we were paying for the place.

First came Averardo, a lofty man of considerable stature by Indian standards, at least six foot tall, he approached as I was out the front inspecting the TV cable that had been 'installed' by draping it across the hefty red silk cotton tree and the undersized custard apple tree in the garden. It appeared to be tapped directly into the left hand side neighbour's house.

"I see you're admiring the workmanship," he offered.

"Yes, it does look somewhat haphazard, doesn't it?"

"I'm Averardo, I live across in these Apartments with my wife, June." He extended his hand.

"Hello, I'm Mick, you probably know we've just moved in. We're renting the place for a few months." I shook his hand.

"Yes, I spoke to Arnold your landlord, not that he likes to talk to me much. He's been building this place for about two years. It does look nice, how much is he charging you?"

I upped the price a bit to Rs10,000, he seemed unimpressed. Clearly he and Arnold did not see eye to eye. I asked him about Arnold.

"He's a funny old stick, used to speak to me quite a bit but hardly gives me the time of day now, I'm not sure what I've done to offend him."

Averardo is an English educated Bombayite. He told me that he moved to Goa when his wife was made redundant because he has family here and it is a more affordable place to live without working. They were missing Bombay though, "Goa is so boring", he advised. He also informed me that he was sixty, he didn't look a day over forty.

"I like to have a drink and mess around, you know, have some fun, these people (motioning toward the neighbour's houses) are all very well but they are so boring, they don't *do* anything."

"Oh, I really like the place, that's why we came here."

"Yes, its great for the likes of you, Britishers with some money in your pocket, but there's nothing for the locals, its all work, work, work during the season and then hibernation during the rains."

I wasn't really sure how to take this, I was hoping for more positive beginnings, more 'here's a cake to welcome you to your new home, don't hesitate to ask if you need anything'.

"Anyway, I'm off for my daily walk, come round to see us, don't hesitate to ask if you need anything."

"Thanks, we will."

Over the next two days all the neighbours came round with cakes to welcome us and offers of help if we needed anything, and of course to find out how much we were paying for the place. Zeta and Pious, Catholics from the house behind. Dinesh and Malini, Hindus from the house to the left, and Marina and Steven, also Catholics from the Apartments opposite.

For the next week we hung around the house familiarising ourselves with our new environment, learning where to take the rubbish (to an open concrete skip down the road where cows came each evening to haul out the rubbish deposited that day and dump it on the road before the dustmen came the next morning to clear it up), where to take the laundry (to Snow White Power Wash where the workers bashed the shit out of it and left it to dry in the blazing sun, leaving it twice its original size and fantastically faded), where to buy our provisions (from Hitesh or one of his three identical brothers in Blackburn Rovers shirts) and where to catch the bus into Mapusa or Panjim, the main towns, a treacherous affair involving standing in front of a speeding bus with an arm outstretched, eyes closed, hoping for the best.

We acquainted ourselves with the flora and fauna. We watched the house crows clash with black drongos in the coconut palms while pariah kites swooped overhead with their familiar 'pweoo' cries. We looked on as sunbirds searched among the copperpod flowers and the parakeets and magpie robins swung on the branches of the same tree, tails flicking. We played with the touch-me-not by the fence, making its leaves close around our fingers and we were happy in the sun and peace. Cows and goats came into the garden all day long, drawn by the scent of the lime tree and the neighbour's banana tree. A ginger cat came into the house and we fed it the remains of our kingfish, pomfret and prawns. At dusk we would laugh at the frenzied cries of the brain fever bird working itself into an anxious stupor and at night we could hear the screeching of the spotted owlet setting off from the cotton tree.

In the evenings we went to the bars to tell lies to the tourists. We would take on different personas, world-weary one minute, travel virgins the next. We had tried to make friends with them but we just didn't find anyone we liked, no one with whom we had anything in common. It seems strange to say this now but they were all too old, too young or too stupid - we had come to the Bournemouth of the East, but we didn't care because it was beautiful. The locals became our friends: the bar owners, the taxi drivers, the shop owners. We were more than *allowed*, we were *accepted*, and it was wonderful.

We found a place to hang out on the beach during the day, a quiet shack away from the throng, run by two sisters and their fisherman husbands and they taught me how to throw a drop-net as we drank King's beer and watched the ghost crabs bolt back to their sand holes. Tourists came and went, stuffed into minibuses, white on the way in, red and brown on the way out. We watched them change colour and leave, but we didn't leave, we stayed. And we didn't work. For a month the only thing we worked on was our tans.

I had replied to the email from my workmate and sent another to Avinash at the Indian company. I had said "Yes, I'm interested", and the reply had been "OK, I'll come back to you", but I hadn't heard anything. I was quite excited about it at first, but in the time between I had become nonchalant, staid. I was happy doing nothing, even though I knew it would be great fun to work here. I told myself I was better off not getting involved, stick to the plan, write that damn book.

In the absence of a telephone (it was finally installed after ten weeks, but was still a twinkle of red tape in a bureaucrat's eye at this juncture) we had to go to the Internet café to access email, our only contact with our former world. The computers were so painfully slow and virus ridden that every visit was a chore. A chore easily avoided by going to the beach instead. After a ten-day absence, responsibility got the better of me and I went, and there, of course, was the mail I didn't want - it was from Avinash, it said:

"Hi Mick, sorry its been a while getting back to you, but we're ready to meet you now, in fact we have been busy with a new market research client and I feel we could really do with your input if you are available. Can you give me a call on the number below when you get this mail so that we can arrange to meet? I hope you're still interested in helping us out. I look forward to speaking to you soon. Regards, Avinash."

I called him and we talked. He seemed a really nice chap and he laid it on thick with a knife - they *needed* me. I spoke to Julie and promised that I would not upset our non-working plans and that if I did get involved and start working it would be on a very short-term basis. She reluctantly agreed. I could go to meet him. The damage was done.

I told myself that I would only get involved if I liked the people, but who was I kidding? I liked *all* the people here. I told myself I would not let it take up too much time, but I knew I'd give it just as much time as it needed. I told myself this was not a career move, just something that came along that took my fancy, but I could see it on my CV already, that CV, the one that I didn't need any more now that I wasn't going to work in an office ever again. My problem is that I am an addict. It shocks me to say it, but the work is the narcotic. A sedative, a downer yes, but addictive all the same. I had been doing so well - slowly making my way through the programme, day by day, in this new world, enjoying the sun, the peace, the serenity, forging a new future with infinite possibilities, taking whatever came along - but it could never last, it was too edgy, too *real*.

One sniff and I was back on the sauce, back on the job. I thought of nothing but the comforting reverie of purpose and the soothing trance of task.

Chapter Five

Inadvertently I had been working in call centres for an indecent amount of time. I started as a 'telephone interviewer' while at college and had been good enough at it to keep the job throughout that time. I must have showed aptitude because fifteen years later I found myself in the same industry, relentlessly researching markets. For many, market research is a parasitic industry existing to augment and reinforce the dominance of the multinational conglomerate by selling it back the views and opinions of its all consuming public. To others market research is a scientific discipline that exists to accurately record and divulge the needs of the public on issues of a social and economical temperament to those institutions in a position to alter such matters accordingly. To most, however, market research is a way to make a living by asking people questions and writing down their answers.

Although market research is a respected corporate career for those involved in deciding what these big companies need to know, for many it is a mundane activity, which entails incessantly roaming the streets, attempting to stop every third person in order to coerce them through a set of inane and exhausting questions about toothpaste. In recent times the telephone has prevailed as the preferred method for gathering such data, which has at least meant that the charge of humdrum inquisition is now largely undertaken indoors. The 'indoors' to which I refer are of course call centres. Call centres come in a diverse array of forms but all are ultimately geared toward the same aspiration - the successful completion of a dull task in the cheapest possible manner. Whether for research, customer service, sales or support, the call centre functions as a factory of dialogue where the drudgery of discourse is succinctly conveyed through a scripted lexis of 'I-need-to-hurry-this-up-sir-I-have-to-make-more-calls-to-keep-on-the-right-side-of-my-supervisor'.

Not every supervisor is a bastard, but most of them are. Not every call centre is a shit place to work, but most of them are. Despite these truths, I had always viewed my position within this world in a quite dissimilar manner to most, the underlying incongruity being that I actually quite *like* the call centre environment. For me the attraction has always been the disparate, desperate, desolate set of unlikely characters that end up in this improbable, inappropriate workplace. I cannot imagine a more contrasting collection of transient, itinerant drifters being allowed to congregate in any other place of work. Resting actors, nascent artists, lazy musicians, dormant politicians, drug addicts, pub addicts, sex addicts, hex addicts, gym addicts, whim addicts - every race, breed, type and creed thrown together in a common purpose. Exactly what that purpose is can very often be, quite understandably, overlooked. Such a blend never becomes stale; the opportunities and possibilities for farce are endless. A few choice examples:

I once witnessed an overtly racist, gay, Iranian ("I'm Persian, I'm Persian") man trying to explain the difference between a guarantee and a warranty to a posh white Rasta who just wasn't getting it. The former (a supervisor) very quickly lost his temper and sacked the latter from his project, calling him a 'Fargin' thick English public school shit-cunt'.

I once caught a French speaking Australian whom I had entrusted with keys (again a supervisor) viewing pornography on my computer at midnight with his sleeping bag and pillow laid out under my desk. He had apparently been sleeping at the office for over a week.

And there's the colleague from Cameroon who, given that his name was Charles Mbantu, sensibly decided to use a pseudonym to introduce himself on the telephone to make it easier for people to understand his name and ensure that questions about his name would not impede his success on the calls. He inexplicably chose the surname 'Creve'. An overheard version of his introductions by anyone passing his desk might go something like:

"Hello, I'm Charles Creve, I'm calling from…"

"That's C.R.E.V.E, Creve"

"No, Creve, not Crab, Creve, C.R.E.V.E, Creve"

"Yes, I'm calling from…"

And there is also my own rather disgusting but most amusing discrimination in the workplace. I confess to always keeping 'pet' staff who are utterly useless and who thoroughly hinder the success of projects but are employed either because they are a good laugh or are so bad that they annoy the shit out of the supervisory staff, which in turn makes me laugh a great deal. One such person, the most hopeless person

I have ever met in my life, who must remain nameless lest she ever finds out the cruel reason for her employment, actually never achieved a single productive working day but was called in to work on nearly every project. Her comprehensive unproductivity was at the despair and utter bemusement of everyone else, but I found their reaction hilarious, indeed, it was not until people started threatening to leave if she was not removed that I relented. As I saw it, she did no real harm. Sometimes it surprises me how seriously people take their work.

Despite the potential for funnies, the call centre job has a shelf life; the unyielding monotony of the work gets to everyone eventually and they move on. A few like myself, move up, take promotions within the field, but most can't wait to get out into their chosen occupations, whether they be gainful or not, and simply leave. No one *really* wants to work in a call centre. It is a means to an end, a clandestine career of which no one is proud, to which no one admits. I rarely tell people what I do, I just say, "I work in an office" and try to leave it at that. If I say I work in market research they always reply "'Are you one of those people who phones me up at home to ask me about toothpaste?"

The call centre looks as though it is reaching its sell-by date in the West. The industry has seen a shift offshore in recent times. In the slipstream of the success of stripping manufacturing overheads and exploiting cheap labour in the East, big business has started to shrug off all of its other unfashionable, unwanted back-office processes and relocate them to the lowest bidder. The market research industry has followed suit and started to offshore its call centre operations to Canada, South America and India. I am still relatively unsure as to how I feel about this. While I have sympathy for the loss of jobs in customer services in the West, I am quite sure that no one I have ever worked with is lamenting the loss of *their* job, good bloody riddance most would say, now they can be free to get on with the business of pursuing a more worthwhile career. The issue on my mind is more one of exploitation - is it constructive and sustainable to offload all the shit jobs to someone else in some far-off land, just because they'll do it for half the price? What will be the upshot of all this? Who gains in the end? I had heard the rumours of nine pounds per month wages in Indian call centres and had taken note of the atrocious conditions in sweatshops in the Far East and I wanted to find out for myself. Here was my chance. As I agreed to meet this company I considered these issues further and tried to rationalise them.

I had read numerous articles in the UK about atrocious working conditions, where call centre staff are expected to work 12 hour days with no breaks, deal with abusive customers, conceal the fact that they are in India, use pseudonyms to further obscure their location and all for around 50 pence per day. Although I had no notion of how much Indian call centre workers were actually paid, all the other 'ghastly' conditions of work seemed very familiar to me. Using a pseudonym is a sensible way to de-personalise the abusive responses, which are encountered on very rare occasions, and while not in any way compulsory, 12-hour days are common among cash hungry operatives in the UK. I have always felt that call centre work is tough, spending all day performing on the telephone is absolutely exhausting, and no one ever said it was supposed to be a nice job.

The vast majority of call centres in which I have personally worked have been in down at heel, low rent, subterranean vaults, with an apparent lack of luxurious amenities. I understood that this was largely not the case in India, and that working environments were usually in large air-conditioned offices with prominent facilities such as fitness gyms, restaurants, relaxation areas, Yoga and meditation rooms, even beauty parlours; a far cry from a scruffy basement in Bermondsey. I also understood that Indian call centre workers were offered pensions, food tokens and free transport to and from work, facilities enjoyed in the position of permanent full time employee. All staff I had ever employed had been on a contract basis with no notice period and until recently no sickness or holiday pay, no perks, and certainly not free transport, again quite different to what is offered to the Indian worker.

I was also concerned about contributing to a decline in British jobs. I have been a member of a union all my working life and felt strongly about protecting jobs in Britain until I began to visit developing countries and witnessed poverty for the first time. I began to consider how much Britain actually *deserves* these jobs. I am reminded of Gandhi's concern over British textile workers when he called for a boycott on foreign cloth, and I feel that same applies here; yes, the unsupported British worker suffers from the job loss, but does this mean India must miss out? In a global society, surely everyone is entitled to a piece of the pie? Why should multinational companies use British workers over Indian ones? Are Indians not entitled to a chance to compete?

My hastily formed opinions encouraged me to look into the job loss issue further, and I discovered that according to a study done by Datamonitor, a market research firm, call centre jobs are actually in a period of growth in the UK and the US, albeit slow when compared to recent years, and jobs in the industry are becoming harder and harder to fill. The area of growth is predominantly in 'high-end' posts, positions requiring a higher level of skill, such as customer relations, which are less successful when outsourced offshore. I have to concur, that in recent years, finding good quality staff in the UK has also been the biggest challenge in my experience.

A common concern raised about the new jobs being created in India, is the lack of prospects for the worker. In order to get a job in a call centre, most Indians need to be at least graduates. In practice, many are far too qualified and in totally unrelated fields. The concern is what happens to the skills these people are dissipating in favour of all this quick cash? Call centres offer very little in the way of career development, and for someone with a biochemistry degree, they offer absolutely nothing. But surely that is the cruel corporate trap where people are tempted into bad jobs with no prospects by the lure of cash? Is this really an argument for limiting such opportunity? It seems tremendously patronising to me to assume that the influx of Western back office work is corrupting the youth of India, surely they should take that decision for themselves?

I therefore decided that if the working conditions were good and the staff were happy then it would be a positive thing for me to aid the process of bringing work and money into India. Indeed, it may even ease my conscience about exploiting India myself - I would be paying taxes if nothing else.

With grand pomposity, I decided to try to view the outsourcing boom as a revenge for the Raj. The British had subjugated and exploited India for so many years, stealing all its wealth and encumbering the growth of industry and endeavour, perhaps now is payback time, where India steals from Britain and beats it at its own game. I always feel a little embarrassed about being British in India, perhaps this could be another way of easing my conscience.

The company was based in Pune and not Mumbai as I had been originally told. Pune (pronounced Poona and was spelt thus by the Britishers) is a medium sized Maharashtran city in the Western Ghats, standing fairly proudly in the shadow of its big brother, Mumbai. Linked by the six lane Mumbai-Pune Expressway, and now a three-hour drive away, the town has burgeoned since its days as a sleepy administration centre under British rule. The current population of somewhere around 4 million are sprawled through the assorted municipal wards, linked by busy trunk roads and leafy boulevards alike. Pune is a green city by any standards and positively emerald by Indian ones. In the centre around MG Road (Mahatma Gandhi Road - every Indian city has one) it could be any Indian conurbation: cramped, crammed, torrid, malodorous, needlessly loud and overflowing with unnecessary hassle. But get out into the Cantonment area or the Boat Club area along the river and the roads widen, traffic thins and the trees take over. Trees are everywhere: colossal, burly banyans with their sinister aerial roots hanging like dead thread; filament arms plunging down to further upset the concrete. Bendy Bhendis line one street, their fruit fruitlessly strewn across the paths like tiny turbans, while Copperpods line the next, constantly in flower, their spindly fronds brushing pollen on to all who pass. Babuls, Ashokas, Amaltas, the odd Tamarind, and figs of all denominations, Pilkhans, Chilkhans, Javas and Jackfruits. Trees engulf Pune in green, making it a quite tolerable place, an almost comfortable place, an Indian city where one might even consider *going for a walk*. But lets not get carried away, Pune is still an Indian city and although flanked by the mountains and penetrated by rivers, it has an industrial past and a technological future. Pune is known as the 'Oxford of the East' due to its high numbers of universities and colleges, and is benefiting from its location and academic prowess. The flourishing Indian IT industry has had its eye on Pune for some time and slowly but surely it is beginning to infiltrate and colonise. Pune is now in direct competition with Bangalore, Mumbai and Gurgaon for the title of Most Outsourceable Indian City. The old industrial areas surrounding the city are now inhabited by Business

Process Outsourcing (BPO) - companies specialising in back-office services for large companies in other countries.

It had been agreed that I would meet up with my potential new boss in Mumbai airport at 8.45am, and that from there we would fly on to Pune together. There are no direct flights from Goa to Pune, although there are direct flights from Pune to Goa, quite how this is possible I had yet to decipher. I ventured into Panjim to book my flights.

I love business travel. I love the fact that I am not paying for it myself and that I don't have to ask for the cheapest option available. I am always very brazen about this when making any payment that I know will be reimbursed, ostentatiously handing over the credit card with an aloof wave of the hand while flagrantly not looking at the price; my own zealous effrontery; playing at being a big shot, such a cheap thrill but one I must enjoy whenever possible now that I am resigned not to work again, now that I am an effusive affiliate of the New Tight. I was looking forward to some business style pampering - business class flights, high grade hotels, choosing the most expensive dish on the menu, looking serious and important, wearing long trousers again. While my new, largely unclothed, beach life was uncomplicated and unflustered, it lacked the couturial opportunities of the working life. Wearing a suit is a prospect that any self confessed, would-be Mod covets, and office work provides plentiful opportunity to wear the three-button, short lapel, six inch vent and ticket pocket. Lounging around on a beach does not. But I had no suit. I had considered a suit chiefly unnecessary in 35° heat and 95% humidity, so I had to compromise - 1960's Ben Sherman, Levi corduroys, Clark's Wallabees, the best I could do.

To reach Mumbai for the connecting flight to Pune at 9am I needed to leave my home in Candolim at 5.30am. This meant booking a taxi and vehemently impressing the importance of the taxi being at my house on time to the driver. While I had used many different drivers in Goa, I had never come across anyone who I considered dependable in any way. It was high time to rectify this shortcoming, I needed my own driver, someone to trust and confide in, someone to guide me through danger to adventure, a Sherpa Tensing, a Rocinante, a Silver. With verve I set out onto the streets to find my man. I was filled with the thrill of possibility, from so many available men I would find my companion, or he would find me, he would come to me through the mist in his gleaming chariot…

There was no one on the street. The gauntlet-run of "Taxi sir? Cheap price. Where you want to go? Taxi? Taxi? Taxi?" Where was it? Where were they all? I searched up and down the main road and found no one, not a soul, not a taxi-man ghost, not a whiff. With furrowed brow I headed home, dejected, broken. Where was my aide and confidant? Where was my sweet chariot?

I first saw him in the distance, perched apologetically on a low wall, staring at the ground, swinging his legs. Barely five feet of timid, subservient, cap-touching Indian man. He had no vehicle but he was sat where taxi drivers usually converged. I approached him.

"Are you a taxi driver?"

He looked up at me startled and turned back over his shoulder, looking for who I might be talking to.

"Me? Yes sir. Where would you like to go?"

"I need to go to the airport at 5.30am tomorrow morning. Will you be able to take me? I live in this house here."

"Yes sir. This will be my pleasure sir."

And so it was that I found my man. He was not quite the cohort I had in mind, not quite the wise escort I had hoped, not at first appearance at least, but I had decided, something was right about him. I asked his name.

"My name is Peter Burgess sir."

Peter Burgess? How can this man be called Peter Burgess? How likely can it be to find and Indian called Peter Burgess? Let alone an Indian taxi driver in scruffy trousers, taxi wallah shirt and chappals? I smiled at him.

"I'm very pleased to meet you Peter Burgess, my name is Mick Sheridan and I will see you outside my house at 5.30am sharp. Thank you Peter Burgess."

I knew he would be on time. He was Peter Burgess.

I went to bed early with the fear of a man who knows he has to rise at 5.30. Even with the alarm set, deep sleep will always be unattainable. The necessity of light sleep so that the alarm is not missed gives rise to a fitfulness known principally among dosser factory shiftworkers, drowsy night watchmen and other on-the-job sleepers. I am a heavy sleeper by nature and under such contrary circumstances am tetchy in the extreme. I lay awake trying to go to sleep. At midnight the dogs woke and put paid to any notions I had about sleep for the night. I got up to drink whisky and throw stones at the dogs.

Sleepless nights are a common phenomena among itinerants in India. The problem is of course noise. It is not people noise or car noise because India usually goes to bed at 11pm. The problem is always dog noise. This is a nightly occurrence in Goa. Packs of wild dogs dash round the streets at night in search of bitches to harass and impregnate. While running they bark and when stationary they howl. The pack will customarily comprise a medley of mutts of diverse lineage, which have become through persistent interbreeding, predominantly small to medium sized, unexceptional, brown dogs without distinguishing features or personalities. They are just dogs. They remind me of slippers in that however much they are decorated with ribbons and bows or are made to look like something else (perhaps footballs or big fluffy pigs), they still remain slippers, and come to think of it, they actually look a little like slippers. Yes, that's it, imagine a pack of ten or more size twenty-five slippers in poor condition running around barking and trying to have sex with female dogs. Now imagine a single German shepherd and a stunted-growth doberman pinscher among the slippers and you have the perfect description of the Goan dog pack.

These dogs are pariahs, exiled by the local community who hate them but ignore them. The local populace is largely resigned to the nightly dog run, and will only venture out to throw stones when the dogs try to have sex with their own pet dogs. All Goans have a pet dog, most are variations of the aforementioned slippers, but some are white and fluffy in the toy-dog style. Toy-dogs excepted, which are walked on leads and doubtless sleep in beds and eat at the dinner table, pet dogs live outside the house and are kept for the sole purpose of barking at each person and vehicle which passes. It is unclear why barking at passers-by is desired by homeowners, but it is wholeheartedly either endorsed, embraced, endured or totally ignored by everyone apart from frightened tourists worried about rabies. One can expect to be barked at as one passes any place of residence in Goa. The residents will be waving and smiling while their dog (or slipper) is baring its teeth and wildly woofing. The nightly dog run has become something of a sporting spectacle in Candolim. As the dogs maraud through the streets, lights flash on in sleeping houses and men rush out to the pile of waiting stones on the verandah. The object of the game is to drive the dog pack past one's own home and on to one's neighbour's by pitching stones at the dogs. This is how I first met two of my neighbours. Mr. Kapoor and Mr. Furtado are adroit players and were less than delighted at my own surprise skill in the game. The first night I played I was a little drunk and had had more than enough of the dogs worrying the stray bitch staying in the derelict house across the road. I tore out of the house and ran at the dogs shouting. No reaction. I picked up a stone and they scattered like fish before I had chance to throw it.

Mr. Kapoor spoke in the darkness, "Silence is needed son, throw first, never shout." I returned to my porch and watched as the pack regrouped. Kapoor pitched – a direct hit, a yowl and the pack ran, looking back with frightened eyes. I gave him the thumbs up as the dogs halted outside Furtado's place. "They'll be back," said Kapoor with subtlety, (no Swarzenegger about it) as he settled back into his chair to survey his pile and select his next missile. Suddenly another yowl and the dogs were off further down the road, Furtado laughed in the darkness, his laughter leading into a hacking cough and ending with the inevitable "hooawk" and spit - the ubiquitous sound of India.

The dogs returned on their cyclical route, driven on past each house until they arrived again at Kapoor's. He lobbed a large projectile and missed, but the dogs got the message and converged on my land. I was ready. I cast a small stone and scored an instant strike, the dogs sped on.

I grew to enjoy this cruel game, such was my hatred for the pack of slippers. I will confess to a number of direct hits too. I must also add that the hypocrisy that is my conscience forced me to contribute a fair sum to the dog rescue centre in Goa. It felt like dropping a five pound note into the plate at church when I hadn't been for a long time, trying to buy my way back into God's good books, paying for salvation - the quickest way to Hell.

The dogs finally relented on my restless night and the whisky offered me a passport to sleep. I woke to my alarm in the darkness. I showered, dressed, tea'd and toasted, took up my luggage and peered out to see Peter Burgess' modest white Maruti taxi van parked patiently and quietly outside my house.

"Good morning Peter", I said as he took my new flight case on which I had splurged a considerable sum to vainly attempt a stylish arrival in spite of the lack of suit. He placed it into the boot with care, "Good morning Sir."

"Please, call me Mick."

"OK Sir."

We drove in silence for a while, it was far too early for conversation. Around us Goa was waking up. It was late October and there was a pinch in the morning air. I was still wearing short sleeves - it was only a pinch. Peter's van wound its way towards the Mandovi bridge through the roadside villages of Nerul, Verem and Betim. I noted Peter's relaxed driving style and felt happy with my choice of driver, I decided there and then that I would persuade him to be *my* driver. I laughed to see people on scooters wearing balaclavas, I pointed them out to Peter, noting also their short sleeves.

"Monkey hats sir, we call it monkey hats."

I laughed, "Please, call me Mick, not sir."

"OK Sir."

As we were crossing the Zuari estuary on the airport approach road Peter started to wail a little and pointed to something on the road in the distance.

"Oh dear, oh no, look sir, she is in the road."

As I looked through the windscreen, past a dashboard cluttered with plastic catholic regalia, I briefly caught a glimpse of something white and round in the middle of the road as we swerved round it.

"What was it Peter?"

"Chicken sir, fallen from rickshaw going to market, her legs are tied. Still alive."

I wanted to go back and help, I really did, the image still haunts me a little. But I had to get to work...

We arrived at the airport and I paid Peter a handsome sum for his trouble.

"This is not necessary sir, 500 only."

"Take it man, you have been up since 5 O'clock this morning."

"Thank you, I have no alarm clock, I did not sleep in the night."

I reached further into my pocket.

I met Avinash at the Mumbai airport exit after nearly thirty minutes of waiting and calling him. A pattern I would have to learn to endure over time.

India does not start on time, at least Indians don't. It is known jokingly as I.S.T. - Indian Standard Time - no joke there, except that this means at least 30 minutes later than I.S.T., which is supposed to be the joke. I confess that I don't find the joke very funny.

I knew as soon as I met Avinash that I liked him. About thirty-five, medium build, medium smile, mild mannered but strong-minded. A very intelligent man, educated at IIM, the Indian Cambridge or Yale, both parents professors. As we flew to Pune together, a 25-minute flight from Mumbai, we discussed the project in hand.

The intention was to build a market research call centre in Pune to collect data for market research companies in the UK, US and Australia and he wanted me to consult on the setting up of the company. He told me that my role would be to oversee the whole thing and ensure that it was done properly. He expressed vehemently that my experience would be invaluable. He dejectedly reported that they realised that with no experience they couldn't possibly do this properly, that they needed me to guide them, to help them be a success, to lead them. He beseeched me to help, earnestly declaring that without me they couldn't do this, they needed me, they were *relying* on me. I am such a sucker for flattery, and despite the Shakespearean ham, I was sold - lock, stock, sinker, hooks, barrel, lines, everything. And I was sold to the lowest bidder - he dolefully disclosed that they couldn't afford to pay me anything like the salary to which I was accustomed because this was India and the money *just wasn't there.*

Before meeting Avinash, I had told myself that I would only take the job if I really wanted it and that I could not be bought for money, not under any circumstances. This was now being very much confirmed, just not in the way I would have liked.

We were met at the airport by a driver with a placard with my name on it. So, they were turning on the style, I would have to be on my guard.

We arrived at the office, an ersatz art-deco slab, a huge cherry granite and marble behemoth, powerful and impressive, a far cry from the scruffy offices of home. I was escorted around the various projects or processes being carried out. It all looked tremendously efficient and impressive. It all seemed so organised and systematic that I allowed my resolve to be broken. My guard dropped. I began telling myself that getting involved with this would be easy and fun, *and* it would help pay for me to stay in India. We had a number of meetings with important sounding people with important sounding acronymic job titles - AVP, SDM, ATM, VP, ASDM - none of these meant anything to me and no-one attempted to unscramble them, their mystery adding to the strangeness and lure. I found myself secretly writing down people's names in my notebook as I met them, as it is so difficult to remember or even understand people's names when they are introduced. Indians tend to speak English pretty quickly, but speed up further when saying names or places, the very words that are unfamiliar; the ones I most need them to slow down on. Everyone I met made it abundantly clear to me that with my input this project would be a roaring success, that this was a great company to work for, that they were all very happy here and that if I joined them we would conquer the world. The gusto was impossible to ignore, and by early evening it was pretty clear to me that I would be working with these people in some shape or form. My expectations had been greatly surpassed, the staff seemed enormously enthusiastic compared to their London counterparts and the company seemed far more professional than I had imagined. If I am honest, I expected it to be a fairly shoddy office with quite basic amenities. I felt a little embarrassed about my narrow-minded deliberations now that my clichéd projection on the Indian workplace had been proved entirely wrong. I remembered a conversation I had had with friends after my first trip to India some years back, a conversation that went along the lines of, "Imagine *working* in India, just imagine what *that* would be like," well, here it was, *this* is what it was like, and it looked pretty ok to me.

I eventually began to display symptoms of fatigue, so Avinash suggested I go to my hotel and that we meet there for dinner later to discuss our future coalition - I gratefully agreed and traipsed off with my little executive trolley in tow. The hotel was called the Taj Blue Diamond. I laughed when I was told the name, and was given a puzzled look. " Blue Diamonds - it's what we call Viagra in the UK," I laughed again. The receptionist just looked more bewildered and called for my driver.

I had never stayed in a business hotel in India and was expecting a fairly low-grade affair. Once again, I was very wrong. This was an excellent four-star hotel with the thickest hall carpet I have ever walked on. In the lobby a blind man in a leather stetson and grubby white shirt played a mini-grand piano very badly, leaning on his white stick. When I say badly, he surpassed even Les Dawson, he was just randomly hitting keys at a varying tempo, I surveyed the lobby with a wide grin to gauge the reaction of my fellow residents and was astonished to discern that the other clientele were entirely oblivious. The atrium entrance was pervaded with the random plonkety sound of insanity and the esteemed business company seemed utterly complicit. Maybe this really was the right place for me after all.

As I rode the lift to my room, I became very conscious that the day's events and the night before's dogs had taken their toll, I was weary and haggard. As I opened the door and placed the plastic card in the slot, I saw the bath. I had not even thought about the bath. I had not anticipated the potential availability of a bath in any way. I had not had a bath for eight weeks and as soon as I saw it I ached for it. I filled it and submerged in the bubbles. Before the bath I was still unsure as to how much to commit to working again, I knew I would do some work for this company, that couldn't really be avoided now, I had no excuse, it all depended on how often and for how much. But the bath swung it. The bath changed all that. After the bath I didn't care. Sylvia Plath once said 'There must be quite a few things that a hot bath won't cure, but I don't know many of them'. If working for this company meant that

I would stay in hotels with baths, then this meant that I would have to work for this company.

Avinash called me to cancel dinner, citing too much work as his pretext, I didn't argue. We arranged to meet for breakfast and discuss terms in the morning. I got back in the bath.

We met and decided on my fate. He could obviously tell I was a sucker for work, and held me captive in the palm of his hand. We agreed a daily consultancy fee and agreed that I would make fortnightly visits for two days at a stretch for the next two months. After this I would spend a three-week period in Pune as the first projects got under way, and we would agree a further schedule after this. It was agreed that Julie could accompany me during any long stays, and that all expenses would be paid by the company: flights, hotels and food.

As I flew back to Goa, contemplating my new job, I felt both elated and embarrassed by my decision to take it on. I was elated and gratified by the adulation and magnitude of worth bestowed upon me by this company and these people who so obviously *needed* me. I was clearly enchanted by the prospect of my own importance. This was also precisely my embarrassment. I had proved my own fickle nature; demonstrated my dependence on felicitation and flattery and consequently I felt something of a wretch. I had come to India and got a *job*, the antithesis of my intent. What sort of fool would *work* when he didn't have to? My sort of fool.

As I drew up to our funny looking house in Peter's taxi, and Julie came out on the verandah to greet me I was awash with guilt. She had left her career as a *paid* practicing artist in London so that we could share our non-work dream, and I had blown it. I had fallen at the first hurdle, just eight weeks in. I felt tarnished and soiled, polluted by my spinelessness. Julie appeared stoic and resolute before me, she was sticking to the plan, she was playing by the rules, she was keeping her part of the bargain. As she approached I saw her pious indignation as holy, consecrated, angelic. Once again I felt so damn *fickle*.

"So, did you take the job?"

"Err...yes, sort of."

Julie looked at me in such a way that I knew she was really looking at me, looking at every aspect of me at once. Her look carried no expression, which of course is the greatest expression of all. She just looked and looked. I shifted on to the back foot. I drew breath.

"Look, it's not a lot of work and the place is great, it is such an opportunity to experience something quite unexpected. The money is pretty good and I don't have any responsibility. I get paid all expenses and get to stay in five star hotels. You can come with me when I travel there and the city looks pretty good, or you can stay in the hotel while I work, they will pay your expenses. The hotels are great; gym, swimming pool, room service, beauty salon, baths."

Her eyes changed. They remained fixed upon me but there was definite widening. I saw inquiry, possibility, even hope.

"Baths you say?.. With plenty of hot water?.. Enough to fill right to the top?"

Fickle? Her and me both.

Chapter Six

It would be three weeks before my first proper visit to Pune for work. Three weeks to get on with the life I had intended to be leading. The first thing to get on with would be the business of acquiring a vehicle.

Driving in India requires fortitude, and for most visitors driving themselves around is not a considered option. Indian roads are notoriously perilous. The absolute lack of discipline, lack of patience and lack of caution prevalent among all drivers guarantees this. Indian vehicles are in worse condition than anywhere else in the world. There are no apparent minimum safety standards that must be adhered to, no compulsory safety features such as wing mirrors, lights, indicators or working brakes. Seat belts are way off the agenda; indeed, one should not take for granted the guaranteed presence of seats. Apparently though, vehicle controls do exist, and the laws are changing rapidly - auto rickshaws in many cities have been compulsorily converted to compressed petroleum gas to reduce air pollution and crash helmets are now obligatory in most states. Emission levels are tested in vehicles over fifteen years old and third party insurance is now required for nearly all motorised vehicles. The enforcement of these controls however, is either entirely fictional or entirely fraudulent. If one is caught without a protective helmet, without a driving licence, an emissions sticker or an insurance document, the police, who do regular random roadside checks, will directly ask for money. Providing you have the required amount (rarely more than 500 rupees and usually no more than 100) you simply hand it over and go on your way. In the event that you refuse to line the policeman's pocket on grounds of principle or simply do not have the money to pay, you will be either summoned to court for a hefty fine and probable loss of licence or simply let off with a severe roadside bollocking that may or may not include a sharp rap with a *lathi* and a broken tail light.

The Indian driving test is legendary. Most drivers have never taken a test because a fraudulent driving licence only costs around 1000 rupees. For the upright citizen, however, there is always the option of taking the test and enduring the three-month wait for the licence to arrive. The test is fantastically simple, it entails demonstrating sufficient control of the vehicle, ability to turn left and right and ability to stop. Use of the horn must also be demonstrated .The driver is unaccompanied and the test comprises an 'instructor' watching the driver going up the road until nearly out of view, turning round and driving back again. Providing the driver can do this without crashing, he has passed.

Indians use the horn incessantly, but the horn is rarely used in anger or in times of danger, it is instead used to alert the driver in front that one is attempting to overtake. Most large vehicles carry a painted instruction on the rear bumper that affirms 'Horn OK Please'. This means 'Please sound your horn to let me know you are behind me'. In most countries people use their eyes to detect following vehicles by the use of mirrors. In India, drivers use their eyes to see ahead and their ears to hear behind. While most people drive with their eyes, Indians drive with their ears. The unofficial rule of the road is 'worry about what the person in front of you is doing and the person behind will worry about what you are doing'. This is the single most dangerous decree of the Indian road. It extends to pulling out into moving traffic without looking, and so shifting the responsibility to oncoming traffic to get out of the way as vehicles stray wildly into their path at a much slower speed. I once asked a hotel cleaner what had happened to his newly grazed and bruised face and he informed me that he had been knocked off his motorcycle while pulling on to a main road. He had obviously not been wearing a crash helmet, and I asked him why.

"If I wear it then I cannot hear the horns of the cars behind me."

India drives on the left, a direct hangover from British rule. India also overtakes on the left, drives on the right, ignores one way systems, traffic lights, pedestrian crossings, road signs and signals from other drivers. India reverses too fast, overtakes too slow, never indicates, always overtakes and doesn't know how to hesitate. Ostensibly, India goes where it wants, when it wants, how it wants, and if one wants to go somewhere too, one better start going along with it.

It was in this spirit that Julie finally gave in to my unremitting and unswerving requests to get our own car. Julie was adamant that we didn't need a car, that we could take taxis or rickshaws to the shops or whenever we wanted to go out for the day, and the rest of the time we could progress on foot. After a few weeks of pedestrian dicing with death and enduring incessant demands to take taxis and look in souvenir shops each time we walked along the road, we surrendered and took the decision to get our own vehicle. My theory to help persuade Julie was that, yes, the roads are certainly dangerous, but due to the impatience of all road users constantly trying to pass the vehicle in front, and the indiscipline of traffic wandering all over the road, traffic flow is greatly impeded and vehicles are predominantly slowed, so while accidents are likely, they are not likely to be *that bad*. In addition, I ventured that although I am an inexperienced driver on Indian roads, and although I am without doubt quite a *bad* driver on any roads, I am still a *better, more disciplined* driver than pretty much everyone else on the sub-continent. How I managed to slip that one by her I will never know.

We first toyed with the idea of getting a scooter. As a self confessed Mod, I had always fancied a Vespa or Lambretta, but had never really had the wherewithal to get one. Price and availability had also featured in the equation back in Blighty. In India, such scooters are omnipresent and affordable. The Lambretta plant was sold to Scooters India Limited in the early 1970s when Innocenti ran into financial difficulties and reassembled in India. Scooters were made to exactly the same template for many years and are still a common feature on today's Indian roads. The other classic Italian scooter, the Piaggio Vespa (meaning 'wasp', after the original body shape) was also produced in the 1970's Indian market and is still produced to similar designs today. Although assembled and marketed under the Indian manufacturing name Bajaj, the scooters were built to cut-outs of the Italian design, the legacy of which is clearly responsible for the omnipresent Bajaj Auto rickshaw, the unnatural extension of the simple steel body and chassis from two-wheeler style icon to three wheel utility passenger vehicle. This could only happen in India.

In addition to scooters, there is also a plethora of old British motorcycles, the type that 'Rockers' would have driven in the 1960s, the most abundant of which is the Enfield Bullet. The availability of these enormous beasts lures Enfield enthusiasts from all over the world, long haired, tanned, road weary foreigners astride gargantuan motorcycles, laden with provisions are a common sight. Similarly, Vespa and

Lambretta enthusiasts locate their aspirational quarry within India, conversely though, they export them back to their respective countries, as I have mentioned, India is not really the place for the style conscious modernist. Nevertheless, the idea of posing on a classic Lambretta, an isolated Mod in a town full of Rockers, appealed greatly to my sense of idiosyncratic self-importance, the only problem was where to get one and how to ride one.

I had never ridden a motorcycle with gears in my life and I found the prospect somewhat daunting. I had driven a three wheel auto rickshaw a few times (strongly recommended, just ask the driver, they're usually happy to give you a go, particularly if you tip them a little something) and had found the gears a real hassle, and I had only ridden automatic scooters in the slightly reticent, tender footed, overtly dangerous and embarrassingly inept tourist way that most people experience on the second day of their package holiday and attempt to overlook thereafter. I asked Peter to lend me his rev-and-go scooter to see if my own two wheel manoeuvrability was any better than that of the three puffy white English girls whose amateur motorcycle choreography I had witnessed terrifying other road users in Calangute that morning. I figured an assessment of my abilities would be prudent prior to splurging out on a two-wheeler.

Peter came round on the bike and showed me how to start it. It had no gears, but a clutch was nonetheless necessary to enable the electric starting mechanism. After a couple of false starts, I finally sped off in lunging spurts towards the end of the road, but as my passage was abruptly curtailed by a protruding tree root, I became abundantly aware that simply not revving was not the same as stopping. While beginning to skid on the sandy road, I pulled hopelessly at the clutch lever thinking it was the brake and as a direct consequence, I witnessed with alarming clarity the tree trunk standing firm and upright and solid and painful before me. Anyone else would revoke the offer of the bike loan there and then, but Indians are less concerned with permanence than the rest of us, and Peter knew that when his bike's time was up, its time was up and there was nothing he could do to influence that fact. Peter the Gentleman picked me up first and only when he was sure I was ok, he picked up the bike. No damage done to either party, and with Peter's non-derisive nature, not even the expected bruised ego. At Peter's insistence I took the bike for a further spin, and eventually I got the hang of things, having spent at least half an hour riding around, even venturing to turn both left and right. The next step in my motorcycle scholarship was to carry a passenger. Julie had not witnessed the aforementioned mishap, so was easily lulled onto the back of the bike. We arranged a trip to Coco Beach.

I sat with the clutch open revving the bike as Julie jumped on the back. The sudden weight and volatility of my cargo was unforeseen but I held firm with both feet, gingerly released the clutch and we lurched off on our two-wheel adventure. This was not actually our inaugural bi-wheeled excursion, we had used a similar scooter in Goa seven or eight years before, when we were considerably younger and less cautious, and if memory serves, it was on this occasion that I hit an unannounced speed bump at about 30 miles per hour leaving us both strewn in the roadside sand.

Everything was going fine as long as we didn't have to stop or turn. It was in these situations that the unpredictable traffic, shifting pillion weight and my inherent lack of balance conspired to make me nervous and wobbly. As we edged clumsily and anxiously through a host of holidaymakers, I could sense Julie's embarrassment at my unskilled steering abilities. Slowly I got the hang of things and my confidence grew, that is, until we experienced a steep incline. A 125cc, even without gears, is still a fairly powerful machine, but leaving the acceleration too late to tackle an abrupt vertical is catastrophic when one has no gear to change down to. After what seemed like an epoch in a treacherously wavering, semi vertical slow motion stillness, we finally fell slowly and silently off the bike. Julie picked herself up and gave me a stare. We agreed to meet at the top of the hill and Julie set off walking while I rode to the bottom and attempted the climb again. I had a point to prove. I fell off again. I walked beside the bike revving slowly, and despite a comprehensive lack of control, eventually got to the top. When we finally reached our destination we were both greatly relieved to get off, but as we walked to the beach we both knew we could not relax knowing that we had to drive the two wheeled bastard home. After no more than an hour we packed up and headed back to our fate. The bike wouldn't start. Julie's humiliation and irritation reared as I looked to her for help. I am a very bad mechanic, I am also a very bad skydiver and a very bad neurosurgeon, that is to say, I am not a mechanic at all. We were at a deserted beach in the middle of nowhere, but fortunately, even India's unpopulated areas are populated. I sought help by flagging down two boys passing on pushbikes. As I have mentioned, one of the most wonderful things about Indians is their almost unanimous propensity and readiness to help others. Ask for help and you will get it from virtually everyone. The boys stopped and kick started the scooter with little effort, remounted and rode off. Julie and I did the same but with considerably less aplomb.

As we skidded to a stop opposite Peter's taxi rank he came across the road looking slightly worried.

"All is ok no sir? I saw you, you looked angry, going like this,' He performed a mime of a terrified person holding handlebars, wobbling all over the place."

"Yes, we're fine, really enjoyed it, no problem, no problem at all. We probably won't borrow it again though, but thanks, we had a very nice day.'

As we walked off I confided in Julie, "Actually, as much as I enjoyed riding that bike, I really think a car is more our style, don't you think?"

She agreed, "Four wheels good, two wheels bad."

The end of my Mod dream.

With Julie's Orwellian axiom in mind we agreed we'd find a car. As you will probably have realised by now, and possibly with some exasperation, in most of my decisions in life I rarely take the sensible, simple option, preferring instead to doggedly and stubbornly opt for that which is richer of opportunity. In this vein, and as something of a classic car enthusiast I decided we must have an *old* vehicle. We would need to hire a car, but a new, reliable and uninteresting means of transportation just would not do. I would not be seen in a Maruti Suzuki for all the tea in India. A classic car it would be, the older and more obtuse the better. Julie is a long suffering and somewhat reluctant advocate of my obsessions, having to put up with a catalogue of uncomfortable, unreliable, unwieldy vehicles on which I have insisted over the past ten years or so. In fairness to myself, she has always finally backed my decisions, and has thoroughly enjoyed nearly all of the 1970's cars we have owned, but I know she would secretly like to have a car that was unlikely to break down on every journey or at least be with someone who had some idea how to fix such a vehicle when it did break down. But again Julie loyally agreed that we should seek out an older, more charismatic vehicle than any of the abundant, uninteresting, tediously ordinary automobiles we had been offered to take on exorbitant rental terms by pretty much every waiter, taxi driver and shopkeeper in town.

There are two predominant 'classic' cars that are still very much available on the Indian market. While there are a handful of sought-after classics dating back to the 1950's, such as Chryslers, Triumphs, and several enormous American sedans, the two most common older vehicles are the Hindustan Motors Ambassador and the Premier Autos Limited (Fiat) Padmini.

The Hindustan Ambassador, or "Amby" as it is tenderly known throughout India, has carved an enviable niche for itself among the country's car consumers by appealing to the hearts of the nation. Its sheer ubiquity in India's cities and towns, and its extensive use as an all-purpose vehicle (taxi, parliamentary escort, family car) means it remains prevalent across the entire subcontinent. The most common model is essentially a 1957 Series 3 Morris Oxford that was imported and assembled in Calcutta. The Ambassador is still the Prime Minister's official car, as it has been almost since independence more than 50 years ago.

The other anachronistic Indian motoring archetype is the Fiat 1100, the dominant three box carriage for Bombayites, the vehicle of choice for over 60,000 taxi drivers. The car owes its origins to the late-lamented Fiat 1100 of the mid-1960s. Which was bought into India as the only rival to the Ambassador and some old Standard 8 models (Basically Triumph Heralds which were also flat packed and reassembled in India).

The original contract ensured the model was branded as a Fiat 1100, but eventually Fiat pulled out of the arrangement, and in 1975 Premier Automobiles Ltd purchased the body presses and started the production of the Premier Padmini in its Mumbai factory. Only last year, the manufacturer pulled the plug on the petrol engine version, due to declining sales but continues to produce a diesel. A brand new 1970's classic.

Both of these cars were an option. A few years back I had seen a European man driving a 'convertible' Ambassador. I have put the word convertible into inverted commas, because it was in fact converted rather than convertible. In short, the roof had simply been sawn off. When I saw it, I could not believe my eyes and shamelessly pursued the man until, through irritation, he finally stopped. I apologised for interrupting him in this most impolite manner and explained that I was most impressed with his vehicle and needed to know how the thing was held together without the most integral component of vehicle infrastructure, the roof. The man smiled, he plainly recognised a fellow incurable car fool-enthusiast, and he opened the door to reveal with some pleasure the reinforced steel rods that had been welded along the inside of the chassis.

"No safety certificates needed here", he beamed. "I just got a local welder to do it, you can get anything done for a little cash."

I eyed the roughly finished chop with awe, the blatant disregard for cosmetic finishing made this a true rat-car, and I loved the attitude. There was no sign of a detachable hood. "What do you do when it rains?"

He fixed me with a stern gaze, "It doesn't rain."

Unfortunately Goa is very short on Ambassadors, those cars that are on the roads are often actually carrying ambassadors, or at least military chiefs and the plethora of other Indian official hangers on. All enquiries as to where one might purchase an Ambassador were laughed off, no one would take me seriously. Getting people to understand that I like Ambassadors because they are short, fat, toad-like ugly and thus *cool* proved impossible, my inquiries were met with bewilderment. It seems that the primary condition for selection of a vehicle in India is mileage. The first question anyone asks when they look at a car or motorcycle is 'how far will it go on one litre of petrol?' Aesthetics, comfort, safety and speed are out the window as far as the Indian automobile consumer is concerned. Goan drivers in particular, know (and will categorically tell you) exactly what they get to the litre, and they will also tell you that an Ambassador has very poor fuel economy and is thus not worth considering as a mode of transport under any circumstances, never mind that it is inexpensive, powerful, comfortable, reliable and safe, and is used as a taxi in many other states as a result. It is just too much of a gas-guzzler for Goa.

With the Ambassador out of the running, my attention switched to the Fiat. The 1100cc engine is permitted among Goan value for money extremists, and the Fiat is *available*. I enlisted the assistance of my faithful driver and consultant on all things motor driven, Peter Burgess, and we got down to finding an outmoded mode of transport. Peter took me to a number second hand car dealers, and each one in turn simply took one look at me and my pasty complexion, sized up my wallet in their minds and endeavored to proffer Jeeps and other such aspirational vehicles. The concept of linking a white man with an *old Fiat* was commercially unfathomable. The next stop was a 'vintage' car dealer who had a few tatty looking Fiats among his largely basket-case wares, but as soon as his avaricious eye clocked my white face, the prices ascended into the exorbitant echelon. Our final effort was to drive around more built up areas and stop each time we saw a parked Fiat, seek its owner and offer a price. Peter's theory was that anyone who owned one of these relics would surely want to sell it and upgrade to a soulless Maruti. He was very wrong. We spoke to many owners who had no intention of parting with their long term companions, and although this meant no car for me, the vehemence and aversion of their rejections gave me great heart, perhaps there is an additional non-value criteria in Indian car ownership after all - sentimentality.

I had been attempting to purchase a vehicle outright on the basis that it would be affordable and cheaper than renting. I had also assumed that renting such a car would not be an option. But now it seemed that buying would not be an option and that renting a disagreeable vehicle would be the only alternative. It was then, exactly as I had surrendered to the inevitability of an *au courant pas* auto that I saw it trundling along in front of us, swaying and bouncing on its way like Postman Pat's van. It was a Fiat, and an archaic model at that, it was battered to hell, and it had no roof. Peter sped after it like a hunter, tracking it with determined concentration, he knew this was the rare animal we had sought all day. As it pulled in to a roadside garage he pounced, swinging the diminutive van in a handbrake turn and screeching to a dramatic stop. We leaped from the van and delimited our target. I spoke to our startled and visibly unnerved quarry in English while Peter stood by, making himself available as Konkani translator.

"Excuse me sir, I'm sorry to bother you, but would this car for sale at all?"

The alarm turned to alacrity and a broad smile spread across the man's somewhat stately countenance, "It is not for sale, but it is for *rent*…at least it will be when it is finished."

This wonderful man explained to me how he had bought the car in Karnataka, had removed the roof and reinforced the body shell with steel rods and scaffold poles (which were clearly visible behind the front bench seat), how he had overhauled the engine and the rest of the running gear and was finally having the thing painted at the garage on whose forecourt we were currently standing, after which it would be available for hire along with the numerous scooters and motorbikes displayed outside his thriving hire business. I could have kissed the man and indeed might have done if he had carried a less upright and dignified demeanor.

After an intimidating test drive which saw me wrestling hopelessly with the dashboard gear shift, kangarooing all over the road and stalling numerous times through poor clutch control, trapping my right knee under the steering column twice, dangerously over steering a corner and reversing into a small muddy ditch, a test drive which would perturb any sensible person, I gripped my new friend firmly by the hand and struck an excellent deal - long term rental at a very reasonable price. The car would be ready in two weeks. I returned to Julie triumphant, she was most pleased with my deal, particularly since it included breakdown cover, an issue about which she had expressed some doubt, knowing only too well my shortcomings in the arena of mechanics.

****

By way of celebration for successfully renting such a great car, we determined to go out and get properly drunk. Whilst Julie and I are both partial to a snifter, especially in celebration, be it of success or failure, I believe that it is safe to pronounce that we are both very positively well behaved when partaking in the joys of drink. And whilst I myself am something of a beer enthusiast, and have been known to drink several pints of beer, one after the other, I am forever at great pains to remain in charge of the beverage and ensure that my guise is, wherever possible, gracious and congenial to those in my vicinity, even when in the throes of vociferous demonstration of my own forthright, unbending principles. This stand is greatly at odds with many of Goa's tourists, particularly among my own compatriots.

Package tourism has altered Goa's laid-back character significantly by inadvertently introducing the older brother of the infamous Lager Lout. The introduction of a new species into an alien environment is usually a recipe for mess, and the Goan atmosphere has been royally messed up by these insatiable predacious interlopers. They are dangerous, unfriendly and largely uninterested in anything other than intoxication. When intoxicated they prey on smaller, more decent holidaymakers and coerce them into agreement that everything about the holiday is bad. I refer, of course, to The Oaf. The Oaf goes on holiday to get drunk, get sunburned and moan about the facilities. The Oaf is between thirty and sixty years old, predominantly male and hails from England, Scotland or Wales. Contrary to popular opinion, the Oaf is never from Ireland, the Irish simply don't moan enough to be Oafs. The Oaf can be encountered in the most surprising places, lured to far off lands by the promise of cheap booze, they turn up in Egypt, Bulgaria, Thailand, Tunisia, Florida, even Morocco, in addition to the more predictable Mediterranean destinations, not forgetting of course, the Algarve, the Oaf Mecca. Oafs no longer go to Spain because there is nothing left to moan about in Spain, it is exactly as they would wish.

Goa attracts more and more Oafs each year. Their numbers have burgeoned in the last three years due to a reduction in airfares. I first witnessed Oafs in Goa in 1999, when staying on a short holiday at the excellent Casablanca Hotel in Candolim. After dinner one night; a tourist barbeque into which we were effortlessly forced, the Oafs arrived. There were four of them, four hideous, cretinous stone-dumb Oafs, four monsters of magnitude. Four fifty year old drunken British bastards.

Patsy, so named for her sallow, pallid but somehow robust skin tones was a Scottish specimen of unbridled ugliness, looking spookily like Viz's Terry Fuckwitt. She knows nothing of life except that it is her individual duty to complain about her lot. Her first drunken words to us were "this place is a fucking shithole."

Her accomplices in stupidity are all unstoppably male. Peter, a fat, burned Southern boor has an unattended crew cut and is apparently her husband although she looks like his grandmother. He can only have come to be married to her through drunkenness. Peter wavers on his chair with his arms aloft and exclaims that he scored two goals against Tottenham in 1963, then changes his mind to 1962. "It seems like yesterday" - a vacant, red faced, distant stare, his eyes opening and closing, his body swaying; a poorly connected appliance flicking on and off, threatening to switch on to standby, sadly denied the requisite current to work for more than a few seconds at a time.

Tommy sits next to him, burping long, slow deliberate burps for everyone to hear, then ostentatiously apologising to the adjacent diners "It's one of life's little pleasures", looking at him it became clear that it might be his only pleasure. He is a five-foot skinny mongrel dog with a fat pedigree pug's face. Silver crew cut. A horrible man.

Ron completes the quartet. A big, fat, sometimes smiling, sometimes bewildered grimacing Londoner, now quieter than the others courtesy of his lost voice. One imagines he lost it through the repetitious shouting of "Tott-ern-am" with which he had been presumably attempting to either goad or encourage Peter, it was not clear which. He is visibly pleased with his huskiness, almost certainly seeing himself as cool Mafioso rather than a plain and simple drunken idiot.

As they arrived at the open-air restaurant, most of the evening's diners had finished eating and were waiting for bills, Julie and I had decided to have a couple of drinks before going to bed. The arrival of the Oafs sealed our decision, we weren't about to miss this. The straggling diners looked about them in horror and then withdrew into themselves, leaving at the earliest opportunity. Being predominantly middle-aged British, they were disposed of the uncomfortable politeness that ensured they lowered their heads and pretended nothing was happening, while the Oafs behaved as if no one else existed; imagine rowdy skinheads performing some menacing horseplay on a packed train of nervous, silent commuters, praying for their stop before anything happens. But the Oafs were clearly too drunk to be dangerous, we could sense this, and waited around in anticipation.

The real comedy began when the shouting Tommy leant too far back on his chair and fell, cracking his head on the wall like a schoolboy in class. He was knocked out flat. Nobody knew what to do. Julie and I were the only diners left apart from a sixty-something couple who were beating a hasty retreat, looking worriedly over their shoulders. Patsy turned to me and asked aggressively if either of us were doctors, I didn't reply. Two tiny waiters attempted to lift the drunken, thirteen stone dead weight back to his chair. Even if they could have got him into it, he would not have been able to sit up. Heeding Patsy's plea, I intervened and rolled him on to his side on the floor, into what I imagined might be the recovery position and inspected the gash on his head. I implored Patsy to stop shouting for an ambulance, explaining that it was unlikely one would be passing a beach-side hotel at 10.30pm, and dispatched the reception boy to call a doctor and tell him that there was a concussed drunk at the Casablanca Hotel who seemed semi-conscious, and that he should come at his leisure.

Ron looked at me and then at the prostrate Tommy and, still sitting in his chair, started to kick him in the chest. He shout-whispered "He's just pissed. He's my best mate. He's alright, he's just pissed."

Moments later, Peter stood up from his chair raising his arms aloft and tried to shout something about football, but lost his balance. He grabbed out for solidity and found only my arm, which is an inadequate buttress for a sixteen stone felled tree. He fell, taking two chairs and a couple of drinks from the table with him. The crash woke Tommy up and while still lying on the floor, he began immediately to sing. He was given a few more friendly chest kicks by his best mate Ron. While Tommy was singing and being kicked, Patsy was patting the blood on his head with a table napkin as the waiters looked on in horror. The opportunist Tommy broke free from Patsy's pats and, lifting himself on to an elbow, made a lunge for Julie's leg, he missed and fell effortlessly again to the ground, laughing lewdly. Patsy turned her pats into slaps and started belting Tommy around his bleeding head shouting, "leave the fucking girl alone you drunk bastard." This symbolised an excellent opportunity for our departure.

We ducked behind a pillar and double-backed towards our room unnoticed. We turned again to view them as the poor doctor arrived - Peter, a flailing beetle on its back, Tommy a bleeding, wriggling worm, Patsy beseeching, desperate on her knees and Ron sitting, laughing, kicking, drinking.

We retired to our room where we could get an excellent view of proceedings without the pain of participation. The doctor, in exemplary Indian fashion was almost completely unperturbed by the drunkenness, obscenity and profanity of the scene. He examined the bloodied Oaf's head with professional calm, heedlessly ignoring the imploring Patsy's dumb questioning "Will he be alright Doctor?"

"There does not seem to be any serious injury, but it is difficult to be sure with head injuries. If there is further pain in the morning I will return. It might be best to stop drinking alcohol..."

"No, he's all fucking right, he's just fucking pissed, he's my fucking mate" Ron added helpfully.

The Doctor examined Ron with a single glance before walking off with his own dignified thoughts. "Yes, I can see that you two might be friends."

Despite the prevalence of Oafs, Goa is truly a most pleasing place to have a holiday. It retains an appropriate mélange of laziness and loafing facilities coupled with the unfamiliar, but protected, 'edge' that the very slightly intrepid package tourist seeks. Indian hospitality is legendary, and Goa effortlessly bestows it visitors with a particularly warm blend of generosity and courtesy.

But is it a place to *live*? Or more specifically is it a suitable place for someone like myself to set up camp? After two months of blissful non-work, I had, in candour, become a little jaded with the plastic 'holiday' environment and started to blame my environment for my own shortcomings and frustrations. I was indeed analogous with the Oafs. Much like the Oafs, I found myself displaying cowardice by directing my outpourings toward any fragile recipient in my vicinity. My main bone of contention was precisely what most tourists actively seek, namely, entertainment, and in particular, music. After two months of searching, I had come across precisely nothing in the way of musical integrity. Categorically nothing any good at all.

Goa is extremely rich in both Indian culture as a whole, and in its own unique contributions to India's mind-boggling cultural tapestry. Music, and predominantly live music, is supposedly of particular significance among Goa's edifying contributions, but I just can't see this, at least I can't see it as significant for me. It seems Goa is not overly enthusiastic about the classical Vedic Hindustani and Carnatic devotional music that preoccupies traditional and intellectual India alike, and is, in short, quite outstanding, but has instead more of a penchant for light hearted, 'fun' oriented compositions of a family entertainment nature. Some of this stuff is very notable, for example, there is the Portuguese influenced pop of Goa's favourite son, Alfred Rose, whose blend of clipped Jim Reeves and Harry Belafonte is revered throughout the state. There is also the Konkani *tiatr* music of Minguel Rod, Alexinho de Candolim and many others whose mix of Konkani lyric with a European sound in trumpets, saxophones, clarinets, banjos, violins and raw acoustic drums reigned supreme through to the 1970's. But these vaudevillian giants are all but gone. With the exception of the occasional Konkani song showcase at a 'dine and dance' night, and the excellent Catholic church service bands, playing typically humdrum hymns, but in an uptempo, unexpectedly uplifting, choppy guitar style, a sort of religious Garage Punk, the dominant musical form in Goa, and indeed among the majority of Western facing middle class India, is Soft Rock.

It is an undisputed fact that Goa spawns more pop and rock bands than any other Indian state, and, make no mistake, these bands are prolific, playing at every opportunity to eager crowds of doting families with a keen ear for a worn out, crushingly slow drum beat and an excruciating sex-lyric, delivered in screwed up face sincerity by youthful hopefuls in bandannas. And this is the good stuff, the *considered* stuff. There are an overabundance of tourist-driven house bands who play the same tired cover versions of John, DeBurgh and the Beatles, night after night to undemanding, unquestioning vacationers who sit silent and applaud dutifully, vacantly after each stultifying rendition. *This* is the problem. There is no discerning audience. Whether it is paying tourists or indiscriminate locals, no one demands more. No one demands better. In fact, it seems to me that it really doesn't occur to anyone to demand anything at all. No one demands because actually, they are happy. They seem to be getting what they want. But it is not what I want, and I began to consider whether I was in the right place after all.

Look, I'm a music snob. I really like music. I don't like most music, but I really like music nevertheless. I have always found that in any given situation one can find music that is at least acceptable. But in Goa I was struggling.

I started out by investigating what was left of Goan Trance. In the late 1980's and early 1990's, Goa became embroiled in the Acid House and Techno dance scene and had spawned its own unperturbed strain of bombed-out beats to drive the hedonistic raves and beach parties being populated by holidaying western drug-hoovers all over Asia. I had caught the very tail end of this in the mid 1990's and much like British dance scene by this time, while the music clearly lacked the intensity and sincerity of its origins, the dogged pursuit and endeavour to replicate the once-wonderful reverie by such a committed assembly of profligates made for a great spectacle. People actively enjoying musical entertainment that they have created themselves is always what I'm looking for, even when it's a bunch of scruffy, worn out old timers desperately striving to rekindle erstwhile hallucinations. Even when the music becomes pastiche, its intentions are still preferable to the manufactured pop pap so beloved of today's youth. Remember, Thou Shalt Not Worship False Gods.

But the Trance scene is almost dead. Killed by curfews, intolerant drug laws and the over demand of baksheesh by greedy cops. The mythic dream is still perpetuated in travel guides and on travel programmes, but the Goan trance rave is today almost limited to one clearly marked 'party space' in Anjuna and the once weekly party coincides with the bloated tourist flea market which Anjuna is so famous for (probably once great, now just tourist crap Made in China, one to avoid at all costs). In addition there is Disco Valley, a moniker harking back to the scene's heyday, an area around Vagator littered with party spaces hosting both legal and illegal raves attended by enthusiastic Israelis, dyed-in-the-wool 'hippy' habitués, tattooed European dog owners, bemused Indian tourists and for a couple of rather bored nights, me. But it wasn't really my thing. Sitting on a blanket, nodding, with glazed eyes really doesn't grab me.

The next stop on my search for melodious distraction had to be the new nightclubs. In recent years, a number of clubs have opened along Goa's beach belt, serving both the domestic Indian and overseas holidaymaker simultaneously. They are well advertised, well organised and well boring. The music swings between Indo-Euro pop and Anglo-American sugar-soft rock. These are the sort of places you can bring your parents, and many do, in fact, the vast majority of attendees, the intermittent young Europeans notwithstanding, are indeed parents, and not recent ones either. These are corporate entertainment nightspots, sanitised beyond interest. They are equivalent to the business conference disco.

The alternatives to these clubs are the Dine 'n' Dance specials for Goan mums and dads. These are hideous entertainment nights held at various specially constructed outdoor venues throughout Goa, venues that are used exclusively for weddings and these community-charity proceedings. Such an evening would typically include a bland Indian buffet meal which you take to your shared table along with your one free beer or port wine beyond which very few people drink. You'll witness some traditional Indian dance performed by reluctant school kids, followed by other school based entertainment such as gymnastics or, if you're lucky as I once was, child-human pyramids. The evening will culminate in another local teenage saccharine American rock cliché in faded denim and headscarves. All this will be tied together by an irksome Master of Ceremonies with a very loud microphone and an eye on the top spot at the Lyons Club or Round Table. His responsibility for the evening will be to repeatedly remind all comers that having a good time involves watching discomforting entertainment with a fixed grin, cheering in response to his gee-up calls and giving ovation after ovation to fat, crooked, local dignitaries and business moguls for their support and community spirit. Not really the sort of night out I'm looking for.

In contrast to family entertainment, there are the city-based nightclubs that are the principal domain of the young and fashionable Indian set. These clubs are much like any other club in any other place, indeed, they are a physical replica of every club in every pop video on every MTV station in every country in the world. An antiseptic blend of sterile social establishment and germ-free adolescents combining to make for the most contrived, un-spontaneous and artless night out anyone could wish for. However, being Indian, there is of course a twist. The music is predominantly Hindi *filmi* pop, which makes for much more rewarding listening and dancing opportunities, while the coy reticence of the middle class Indian youth to indulge in overt sexual behaviour pitched against the underlying desire of every youth to actually engage in the aforementioned sexual behaviour, makes for an exciting if frustrating atmosphere, sexually charged with lots of pouting and posing, but you'll be going home alone. So far this was as close as I had come to finding entertainment, but it was still crap, at least for someone of my age. The music played is the same every night, and the whole youthful exuberance and charged atmosphere is too much for a grumpy middle-aged old soak like myself. I suppose the average age at such a club would be around 22, so I didn't feel granddad old, not shamefully old, but the preordained playlisting of Bryan Adams' 'Summer of '69', played at least twice, every single night without fail, is frankly too much to bear, in fact I am uncomfortable even thinking about it, the way everyone would start whooping and wielding air guitars... Christ, this revolting depiction must stop now, I think the point has been made.

Back on the hunt for aural stimulation, I decided to embark upon a full tour of the beach shacks from Baga right down to Aguada. The beach shacks in Goa are coconut palm huts constructed along the entire beach in November and deconstructed in June before the monsoon winds get a chance to tear them apart. They originated as disparate oases of shade where the charred beach wallah could escape the uncompromising penetration of the midday sun, have something home cooked to eat and drink a quiet beer with similar minded sojourners. Today the beach shack is an ardently contested business where large sums of money change hands with dishonest officials for the dispensation to vie for a small slice of the none-too-huge tourist money pie. The upshot is that the beaches are lined with an unbroken terrace of these huts, of varying size and facility, that are either totally devoid of human life, or bursting with bulky, blistering tourist hides. A tourist trait the world over is to flock with one's own breed, and when faced with a suspiciously empty eatery or an uncomfortably full one, most will opt for the latter. The result is a vigorous contention between shack owners to attract this human livestock to their pen. Sun beds, familiar or comedy names, pool tables, surf boards, barbeques, fishing rods, beer promotions, fireworks, all are frequently used as tourist bait, as is, of course, music. Getting the music right should be the key to successful entrapment, but amazingly, it seems to have no bearing on the clientele whatsoever. In most environments where music is played, one does not need to use too much of one's brain to surmise who will frequent that place. A Country and Western night will attract Country and Western fans, a UK Garage night will lure people wearing Reebok Classics, an opera will draw people in tuxedos, a gig by The Fall will yield an extraordinary mix of uncommon characters, united by their own disharmony. But this beach is the antithesis of common experience. Music is not a factor that either unites or divides this audience, they will just tolerate anything. Absolutely no preference for anything. The Pink Floyd Shack will be as full of old gimmers as it is old stoners. The Bob Marley Shack is as likely to be playing Boyzone as anything else. Only Fools and Horses Shack might be booming out Hindi Pop, which will be switched to Meatloaf as soon as the hapless visitor takes his seat. For me this is most disconcerting. Where is the quality control? How can one exercise one's own social codes and signifiers in the absence of learned rules? The way we dress, the things we like, the places we go; these are the things that define us, but none of them matter here, none of them work. Like a bird without colours or songs, there is nothing to set anyone apart from anyone else, no social groupings, no context.

It took the best part of a week of days and nights to even walk past all of the beach shacks, some 150+ in all (I lost count at every attempt). I found almost no difference between any of them. There were a few playing Indian music, but they were usually empty as a result. There were a couple playing fairly interesting dance music, but it always amounted to one CD that could be changed to No Doubt without any warning. I encountered The Specials as part of an 80's party tape where the next track was Spandau Ballet, I witnessed an excellent British bhangra compilation be succeeded with Enya. In short, I found no consistency anywhere, there was not one place where I could be sure I would not be suddenly perturbed by Shania Twain and have to leave, there was certainly nowhere I could be assured to categorically keep away from Prince.

I once asked the host of my favourite sports bar, Bom Sporto, why they never veered from the well-beaten path of Karaoke and local party bands for their in-house entertainment, and why they didn't try to attract some of the younger crowd by experimenting with some different music, I even offered to organise an 'alternative' disco myself, but his reply was indicative of the Goan approach - Why change anything when it is working? We have to think of our clientele, this is what they want, why take a risk by trying something new?

Exhausted, I headed back to my favourite shack in defeat. My favourite shack is simply called Ro-Max because it is owned by Rose and Maxine. They have no music and no frills, the menu is limited, there is no party atmosphere and they close as early as possible. I sat looking at the silently beautiful beach and gave up on music for a year. As long as I stayed here, there would be no decent music for me, instead I would opt for no music at all. As the shack closed I walked home past the Bom Sporto and popped in to see Anupam. As I walked in to the bar, I encountered the impossible, Jackie Wilson singing 'Nothing But Blue Skies', and at some volume.

"Anupam, where did you get this CD? Who gave you this to play?"

"This man here", he offered, pointing to two fifty year old men playing pool. "Is rubbish no? Very old. Too much of old music this man likes."

Could it be? A northern soul fan in Goa? I confess to getting very excited, sitting very attentively at the bar, watching the man, trying to casually catch his eye, I didn't want to seem too keen as I zealously examined the compilation CD Anupam had handed me. 'Floor Fillers - 45 Classic Northern Soul 45s'. I had my own copy of this very disc with me back at the house. I had to befriend this man, maybe we could start a club night together? Maybe he had even more CDs with him than me and we could trade a few and have them copied? Maybe there was a whole host of soul fans in Goa who met secretly every week to dance, just like they do in the UK? He was around 50 and was certainly from the north of England, almost certainly Lancashire, maybe he was an original Wigan dancer? Maybe he had been into the scene from the very beginning?

Finally the man approached the bar to get another drink, as The Artistics belted out 'Have More Time', I presented the CD case to him.

"I understand this excellent collection is yours?" I proffered, smiling, but not too much, trying to look cool, trying to make a friend.

"Yeah, I found it in my suitcase, it must belong to our kid, I gave it to the barman, I think it's a pile of shit, you can have it if you want."

"Oh. Err…thanks, yeah."

I watched him go back to his miserable pool table with his miserable beer and I miserably gave up again. Music was over, I'd better get used to it.

"Anupam, keep this CD here and play it when you see me walk in, particularly when I look like I'm in a bad mood."

But I knew he wouldn't. Televised sport is the order of the day in this bar, and any rare opportunity Anupam gets to play music, he will play Bryan Adams. And so will everyone else.

****

Defeated by music, I retreated into my plan to visit Pune again by making the necessary preparations. I bought Peter an alarm clock to save him having to stay awake all night again and booked him to take me to the airport at 5am.

A semi sleepless night ensured I was even crankier than my typical morning self and conversation on the airport journey was at a minimum. I arrived to find a queue of British elderlies patiently waiting to check in their luggage for the Cochin flight, presumably a pre-booked Keralan backwater excursion where they would stay in a fine-looking *kettuvallam* house-boat bedecked with reclining cushions, luxurious bedrooms and an on-board chef, and drift through the rivers and canals, lakes and lagoons, relax, unwind and get comprehensively, inordinately and agonisingly bitten by mosquitos. Despite my ill-tempered mood, I spared my queue neighbour the minutiae of my own hideous experience of the same, and merely ventured "Do not get on board if there is no mosquito net, please, I speak from bitter experience." Noting a lack of urgency among the swelling queue, and in growing concern over my own lateness (the flight was due to depart in thirty minutes) I asked my neighbour what time his flight was due to leave, "In two and a half hours."

I raced past the entire queue to the single open check-in desk waving my ticket shouting "Mumbai, Mumbai, sorry, running late" by way of explanation for jumping the queue, something that needs clear explanation among the British if unwanted bilious comments regarding 'why we fought in the War' are to be avoided. While I was checking-in with hand luggage only (always avoid checking in luggage in Indian domestic flights, particularly if you like the look of said luggage), I imparted to the huffy looking blue-rinse at the front of the line who had obviously been up since 3am, "It's really not necessary to check in so far in advance for domestic flights in India, honestly, they use an excellent common sense approach to flying, particularly Indian Airlines, they'll let you check in with about twenty minutes to go, half an hour is ample." And I raced off, feeling one hundred tired British eyes perforating my back.

I arrived in Mumbai and made the connection to Pune without incident. On leaving the airport however, I could not find my allotted driver. It was with great disappointment and some embarrassment that I stood before the crowd of vying corporate collectors thrusting their placards at me nodding enthusiastically. I stood for over a minute scanning their boards animatedly, ducking and raising my head, straining to spot my name on a placard below and above their writhing arms, legs, bodies, heads and noise. Finally I realised I had been forgotten and pulled back inside the airport to call the office.

"Hallo, TDS."
"Hello, can I speak to Sharmila please?"
"Hallo? Hallo? TDS."
"Can I speak to Sharmila please?"
"Err Hallo? Sharmila is not there." Click, Click, Errrrrrrrr…
I called again.
"Hallo, TDS."
"Hello, this is Mick Sheridan, I am at the airport waiting for a…"

"Err Hallo? Hallo? Hallo sir, who is there?"
"Do you speak English?"
"Moment please."
Muffled voices, sound of the phone being plonked on a desk. Finally...
"Hallo, TDS."
"Hello, this is Mick Sheridan, I am at the airport waiting for a driver to pick me up, do you know if he is on his way?"
"Hallo Sir, TDS."
"Oh for Christ's sake, is there...oh, forget it."
I walked over to the pre-paid taxi counter and booked a cab, at least I would get a receipt.
I arrived at the office to find that the driver had picked up another person from the same flight but had not waited for me. I glared at Sharmila, my travel contact; she apologised and gave me my itinerary for the next three days. I was again booked into the Taj Blue Diamond Hotel, but I had meetings practically all day so wouldn't be checking in until evening. I set off to meet Sudesha, the Transition Manager for market research, and my first meeting of the day.
"Mick, how are you? Come, let's go for a smoke."
We descended in the lift to the building's rear entrance where smokers huddle round the doors for a stolen nicotine moment before heading dutifully back to the reality of work, a practice identical the world over. I consider myself a part time smoker, but am in reality just a smoker. It is rare that I smoke during the daytime, and rare that I smoke without a drink, but I make up for this when drinking during the nighttime. I occasionally smoke at work, but usually after having been in a pub at lunchtime. Lunchtime drinking is a sackable offence in India, so I imagined I would not smoke much, but it would be hard to avoid, smoking is used as an excuse to get out of the office, there is really nowhere to go during a lunch hour, and consequently very few managers take one. Instead, they go for cigarettes on an hourly basis and drink tea as an aside. Not healthy, but when was work ever healthy?
We settled the plan for the day while smoking. I would be meeting the management *team* who I would be advising later that day, and would meet the first batch of calling agents, the interviewing *'team'* the next

day. On the third day I would simply fly back to Goa, there would be no need to come to the office. This sounded an excellent start to a very non-involved appointment, and while exactly what I was supposed to do with the respective 'teams' was unclear, I assumed that the role of consultant would entail me deciding such things for myself, so I determined to just wing it and see what I came up with.

I have put the word team in italics because I am slightly embarrassed about it. Working in any connection with the marketing industry necessarily gives rise to the use of marketing speak. After so many years of heated protest, I have finally relented and accepted the inevitability of some of these phrases. 'Team' is one such. I recall working as Saturday boy in a record shop while still at school with a sardonic saxophone player who would cross the threshold each morning, raise his shoulders and sing out 'Morning team' in a mocking tone, while shooting a mordant after smile towards the corporate minded, money grabbing, Capitalist Pig Boss. At least that's how I saw things then. Subsequently I have always slightly winced at the notion of anything that calls itself a team but is wholly unconnected with sport. But this childish compulsion is impossible to maintain in an environment where everyone is industriously *building team spirit*, busy becoming *movers and shakers*, and categorically *putting things to bed*, and although there have been many occasions where I have plainly recoiled at such terms in full view of their protagonists, if one is to accomplish anything in the marketing milieu, one must *toe the corporate line* to some extent.

There are of course boundaries, limits. One must impose equitable sanctions if one is to avoid total immersion in the unlimited world of the marketing metaphor. I categorically never *put anything on a back burner*, I try very hard not to *touch base* with anyone, I refuse to give *ball-park figures*, giving perfectly good estimates instead. I remain active rather than *proactive*, I don't *take things offline*, I insist on using my brain, even when it's *a no-brainer*, and on no account do I ever *add value*. While I can defiantly (and smugly) pronounce the above, I must come clean and declare that on more than one occasion I have slipped up and forgotten my principles, I have *revisited my game plan*, *gone the extra mile*, and on one very extreme occasion, I am ashamed to say that I *thought out of the box*. But on all these occasions I was pushed, I was goaded into compliance by some other perpetrator. I can be proud of my efforts to save this wonderful language from corporate ruin, I only hoped I could remain true when confronted with the Indian version of this despicable infiltrator. I wondered how embroiled in business bullshit my Indian colleagues would be.

I elected to speak to the management team together as a group and then again individually. I wasn't sure what to ask them, but I was resolute that despite my eagerness to befriend these people, (after all, they were almost certainly going to be my only social network while working) I would try to judge their suitability on my own terms - I wanted people who could do the work without me, people who could cope on their own. If they couldn't, I'd have to *make waves*.

Three people filed into the office, Sarasvat, Shanti and Manohar. Sudesha introduced them all as having been specially relocated from Delhi to staff this new venture. A fact I found quite remarkable, particularly when I realised they had little or no understanding of market research, or even had a clear picture of what it was they had actually been employed to do. I knew that market research was new to Indian call centres, but I was surprised at how readily they had relocated to a new town and a new company without any real notion of whether they would be happy in the job.

"So how much do you know about the task we have in front of us? Tell me what you know about the market research process, that is, what we're going to be doing and how we're going to do it?"

This was met with a unanimous blank look. I rephrased the question, elaborating further, asking them how they saw things developing, and was met with more uncomprehending glances and lots more looking at the floor. I began to feel a little concerned at their reluctance to talk, surely they were going to try to impress me? They had come all this way, and it seemed to me quite a lot was at stake.

"OK, talk me through your experience, tell me what you've been doing in business up till now."

Shanti spoke up first.

"Our experience has been in working for a number of call centres in Delhi doing 'voice processes'. I guess we'll be building a team to do market research questionnaires in much the same way."

Sarasvat picked up the thread.

"Actually, none of us have any direct experience of market research itself, but from reading around the subject it is fairly clear that we'll be required to ask questions and record answers, collect data from people. To be honest, compared to hardcore sales or credit card collections, this seems pretty straightforward. I don't want to simplify it too much before actually experiencing it, but between us we have a very comprehensive appreciation of how to go about business in Indian call centres, don't worry, we'll pick it up and we'll make it work."

Manohar finished off.

"See, it's like this, give us a better idea of how things should work and we'll sort the rest out, as long as you give us some guidance on the right procedures, the rest will fall easily into place, I mean, how hard can it be?"

I needn't have worried. I think politeness was holding them back, waiting for me to lead the conversation out of respect, something I'm not really used to. We discussed the course of action, functions, responsibilities and roles we would adopt and I quickly ascertained that their methods were very well suited to the undertaking, and that I would almost certainly be able to leave the whole thing up to them. For a moment I had been somewhat concerned that I might end up doing everything myself, that I might actually have to do some work. I dispensed with the idea of meeting them individually in the office, this clearly wouldn't work, I instead decided to orchestrate a social environment to crack the frostiness.

"OK, that all sounds fine, shall we go out for a drink to discuss the rest? I'm a little tired after a long day, shall we say after work tomorrow?"

This time broad smiles, use of the old international language. We filed out of a room littered with broken ice.

The next day I arrived at the office half an hour late after a number of frustrating phone calls to establish that the driver allotted to pick me up from the hotel had failed to do so, and some more to ensure that a replacement had actually been sent. After some time sitting in the office reception, being stared at by each person passing through, I again met the management team and we proceeded together to the training room to meet the calling agents, the people who had been hired and were being trained to actually do the work.

As we entered the room I felt like a schoolteacher as a semi circle of twenty eager apprentices stood up from their seats to attention and looked me expectantly in the eye. I pondered whether they had been made to sit boy-girl-boy-girl as I tried to make sense of the alternating rainbow before me. The men were in typical business drab or 'formals', while the women shone in a resplendent spectrum of *salwar kameez* colour, the 'traditional formals' many women wear to the office.

"Err...please sit down, there really is no need to stand when I enter the room. My name is Mick Sheridan, as you will doubtless be able to detect from your own voice and accent training, I am not Indian, I am from England or as you might say 'I am from UK'.

"I have been working in market research for many years and have come to help and advise on how to make your task most successful. I will help you in training and I will advise the management on how best to go about this business, which is very new to India. I understand that you have been in training for six weeks already which I must say surprises me somewhat, in the UK I would typically give people one or two days training, but they of course don't have to go through voice and accent training. From what I understand, you have been through instruction on accent neutralisation rather than being taught to speak in an English accent, is this true?"

This was met with a unanimous and resounding "Yes."

"Good, because there is no such thing as an English accent anyway, and while it might be very amusing for me to hear you all attempt such a thing, I am very confident that it would make an embarrassing mockery of the work we are here to do. It should be very clear that we are not actors, we are not attempting to mimic the work processes that exist in other countries, we are trying to replicate or mirror them as best we can, while maintaining our integrity and honesty which will be hence reflected in the work we do. Before I take questions from you all, I want to ask a few of my own to ascertain how much you know about what you are going to be doing."

I asked how much they knew about market research and how much preparation they'd had for the job they would be doing. It transpired they knew absolutely nothing about either. They had no idea what they would be doing other than it would be making or taking calls. They had all taken a full time job without the slightest notion of what it involved.

"So no-one here knows what he or she has been employed to do?"

"Well it is call centre no? We will be taking calls from customers isn't it?" Offered Chandrak, the most brazen and nonchalant in the group.

"No, you will be making calls to people to ask questions about a number of things, toothpaste mainly."

"Toothpaste? We will be working for toothpaste?" requested Priya somewhat doubtfully.

"In a roundabout way, very probably" I retorted snappily, angry at how uninformed they all seemed.

After some time I extracted that all company staff are employed with a certain project in mind, but go through a training period of 'voice and accent' training to ensure they can communicate well and 'cultural training' to ensure they understand the culture they are attempting to communicate with. After this it is revealed to them exactly what their job is and the 'process training' starts. It seems the process training had not started yet.

"So when does the process training start?" I asked Sudesha who seemed to be the only one who knew anything tangible about the project in hand, for example when it might start.

"The client will do the training when they come in six weeks time."

"Six weeks, what are they going to do in training until then?"

"I don't know actually, we'll think of something."

I arranged for an emergency training session for the next day so that I could at least give them an introduction to their destiny. I wasn't looking forward to it, how do you tell twenty mustard keen graduates that if all goes well, for the next two years they will be asking the Great English Unwashed an unchanging set of questions about the products they use to wash themselves, in an unrelenting and uniformly tedious manner, day in, day out, without respite and without any thanks?

In preparation for this onerous task, I arranged to meet a few of the managers for dinner to get better acquainted with Indian working methods and Indian drinking methods.

I met Sarasvat, Shanti and Manohar in an English theme pub called 1000 Oaks and they told me their stories.

Sarasvat, a six foot two bulk of a man and a positive hulk by Indian standards, was born into a military family and consequently grew up all over India. Relocation from Delhi where he had been working for the past four years in another call centre was no problem although the food in Pune was troubling his stomach and this troubled his mind, as eating and drinking were clearly an enthusiastic undertaking. He was also missing his new wife who was still working in Delhi in the same call centre and who was considering moving to Pune after a few months if everything worked out for Sarasvat. It seemed his corporate development prospects were good enough to warrant a split, at least in the short term. Sarasvat is a quiet man, unlike his friend and colleague Shanti, who, at 35 had got into the world of work very late, some ten or twelve years after graduating. Being from a particularly wealthy and equally congenial Delhi family had enabled her to spend her youth with the young, fashionable set, concentrating on the louche, lounging Indian activities of "partying hard", "having a blast", and "really freaking out, yaar?" But the time had come to settle to marriage or career; she chose the latter and was doing very well at it in the confident, certain and unfailing way that only good breeding ensures. Now a determined careerist, Shanti demurred her past, hinting at it only through her kitten heels, Gucci bags and Chanel sunglasses, oh, and the omnipresent Cosmopolitan magazine.

Manohar was a colleague of both Sarasvat and Shanti in the same company in Delhi. Having graduated from the Institute of Hotel Management, he aborted his kitchen career on the basis that it was too poorly paid, and embarked upon his call centre vocation determined to atone for lost time. He rose very quickly within his field and was deemed ready by Sarasvat to accompany himself and Shanti in this new venture in Pune. His ability to fastidiously recall things learnt and implement them rather than just making things up as he goes along, makes him an ideal candidate for such a job. This aptitude also enables him to remember everything that he has ever done and recount it in detail, even extending to the contents of each item on a menu right down to the requisite quantity of salt and pepper. For this talent, he has the epithet "Mister Ingredients", a very useful man to have around when ordering food. A very useful man at drinking beer too, as it turned out. It has been my experience that Indians are largely unable for beer, unless of course they have spent time in the UK or Australia. Spirits yes, no problem there, but any number of glasses of beer in succession above four usually culminates very quickly in a short burst of silliness and sequential puking. But not these three. I asked Manohar if he had hollow legs, and being the practical type, he replied that no, he had normal legs and why would I think they were hollow? An exemplary answer after seven bottles of Indian Fosters. I became certain that I could count these three as my friends.

The next day I was again late for the office as a result of the driver who apparently arrived at the right time but did not report to reception, so when the office insisted to me that he had arrived, I went into the car park and asked a man sitting in a large unmarked vehicle if he was from TDS, he simply said 'yes' and started his engine. He had apparently been sitting for half an hour expecting me to go looking for him in the car park.

The new recruits were ready for my crash course in market research, most of them had done some background research themselves and already had a good idea of what it was all about. The theory was very simple for them and they were eager to get on to their main concerns, which were that they would not be understood by the people they were calling. I tried to reassure them that people in the UK are well used to Indian intonations and would not struggle to comprehend them, rather the problem more likely to arise might be that they were unable to understand the people they were speaking to. I fielded all the usual questions like, "Will they be rude to us?" , "What do I do if they won't talk to me?" , "Will they slam the phone down on me?" and tried to reassure them that the entire population of the UK was polite and friendly and sophisticated enough to understand the value of market research and that they were pretty much just sitting there waiting for the call. I tried to get them to concentrate on the rules and techniques.

By early evening I judged that they were ready to make some calls. I announced this and the room froze, "What, today?" they all seemed to say.

"Yes, it seems to me that you've spent the last six weeks becoming more and more scared about the job you will end up doing, and given that until yesterday you didn't even know what that job was going to be, I'm not surprised. By the end of today I want everyone in this room to have completed this simple interview that I drafted this morning, with a complete stranger in the UK. I don't care how good or bad you are at it, I just want you to have done it so that you can spend the rest of your training period focusing on the important issues. You will not know what these are until you actually do some calling. Here are the questions, read through them and when you're ready, go down to the calling floor and phone some random numbers in the UK, Manohar will show you how to do this, as soon as you have done an interview, come straight back to this room."

With this I left the room and proceeded to the calling floor to wait for them. They arrived and started calling. They were hideously under-prepared. The first stumbling block was actually the word 'Hello'. It seems that hello is only used in opening a call in India, and saying hello at any other time is an indication that you can no longer hear the other person. English people of course say hello again after the caller has confirmed who they are. Conversations were going along the lines of:

Ring Ring…"Hello."

"Hallo Sir, my name is Ravi and I'm calling from a market research company.."

"Oh, Hello Ravi."

"Hallo? Hallo? Can you hear me sir?"

"Yes, Hello."

"Hallo? Hallo? Are you there sir"

"Yes I'm here, Hello Ravi, what do you…"

"Hallo? Hallo? Hallo?"

Having sorted this out to at least some degree, we began recording the calls and I went back to the training room to await those who had successfully completed their call.

The first three came in together in an advanced state of excitement. The others sporadically drifted back to the room, and as each of them entered, the others clapped and whooped. The level of exhilaration surprised me, but it was clear that my 'in at the deep end' approach had worked. I was amazed at the level of detail used when recounting their experience to one another and to the group, it was apparent that this had been something of a milestone for them, and I felt much better about them doing what I see as a pretty awful job, maybe they could enjoy it after all.

They went home happy and I went back to the calling floor to listen to the recordings with the managers so that they could focus the training on the right areas. By and large, the recordings were ok, with only the occasional toe-curling exchange, and we discussed a training programme that I would send from Goa and that they would take the new recruits through before the first clients arrived in six weeks time. I told them to focus on listening skills in some way because the main problems had arisen as a result of them not being able to understand the people they were calling, and just before I left that night to return to Goa the next day, a most pertinent example of this cropped up on a recording of an interview with a fine gentleman from Yorkshire.

"That is the end of the interview sir, I just need to ask what is the occupation of the chief income earner in the household?"

"That's me, I work on a farm."

"I'm sorry, I didn't get you, you work on a faam? That's F. A. A. M?"

"A farm, I work on a farm, I'm a dung spreader at the moment."

"Sorry, sir, you said you are a doom spreader at the moment?"

"Christ no, dung, I'm a dung spreader, I spread shit around, I spread shit all over the fields..."

"Sorry sir, you are spreading shit? Yes?"

"Yes, muck spreader, I'm spreading dung."

"OK, is that a manual or clerical job?"

"It's on a farm for Christ's sake, I spread shit on fields."

"I'm sorry sir, what is feeords? I didn't get you."

"FIELDS, FIELDS, oh forget it, I've had enough, I'm off."

"I'm sorry sir, you're roff? What does it mean sir?"

Click click, Rrrrrrrrrr....

"Hallo sir? Hallo? I can't hear you sir, are you there? I just need to record your occupation and the interview is over, are you there?"

Rrrrrrrrrrrrrrrrrrrrrrrrr.

Chapter Seven

Back in Goa, Julie had taken receipt of our new car. Peter collected me from the airport full of squeaky voiced excitement because he had seen Julie driving it that morning and he knew what it looked like. As we pulled up to our curious orange home he was bouncing in his seat like a gibbering, grinning chimp. And there in front of Noddy's house, was Noddy's car. I had managed to deter the car's owner from spraying 'designs' on the bodywork and we had agreed on a classic plain white. But it was still Noddy's car. A tiny strap-on roller-skate of a vehicle with reverse opening front doors, picnic blanket plaid seat covers, a very roughly hacked off roof and an 850cc engine. Julie greeted us smiling at the door and threw me the keys. After yanking haplessly on the adjustment handle on the front bench seat a few times I realised it had been rendered obsolete by the chassis strengthening scaffold poles welded behind it, and I gave in to the inescapable elephant-on-a-moped driving position. We took the road into Calangute, and as we crossed the tiny roundabout to Arpora, kangarooing and demonstrating some unprecedented driving ineptness, a serious looking old man called out from a roadside shop.

"Sir and Madam, I fear your roof is missing."

Being mobile (after half an hour or so I managed to master the erratic clutch despite having to constantly keep my right foot hovering above it to avoid either getting my knee jammed painfully under the steering column or riding the clutch in a graceless, vrooming learner driver style) gave us a whole new take on Goa. We had taken taxis to the various 'sights' on numerous occasions, but we had never really had the freedom to roam and poke our noses into less celebrated spots. We stumbled upon some remarkable locations, some very isolated and tranquil cul de sacs, quite off the beaten track, and principally through getting quite royally lost. There is no such thing as a map of Goa, or indeed India. There are of course numerous pretenders, largely delineated by either tourist touting hoteliers whose establishment predominates an otherwise back-of-a-fag-packet pencil drawing, or by deranged cartographers with a lazy pen and a very mean streak. And it seems that Indians are entirely apathetic about their badly charted terrain. Each remonstrative remark I made about getting lost was met with the ever-present and blithe wave of a hand, *'chalte hai'*, *'ces't la vie'*, *'so it goes'*. Clearly everyone is content to drive for miles in the hope of being pointed roughly in the right direction, relying chiefly upon conjecture and the grunted retorts of questioned roadsiders who, when asked "Which way is Margao?", reply "Graaanph" and point first left, then right, and finally left again before hawking a huge spit inches from the car.

So giving in to the inevitability of disorientation, we eschewed our niggling doubts about the reliability of our sawn-off 1950's home-made sports car and set off on several fairly aimless excursions to have a look round. And we turned up in some staggeringly good places.

We discovered an area called Reis Magos after the Three Wise Men who gave its fourteenth century church and fort their name (or Res Mongoose as the locals strangely refer to it). A sweeping estuary bank road bedecked with stately residences, resplendent with a view of the capital city and tidal lappings on one side and a thoroughly forested mini-mountain on the other. It was here that we saw our first python, only two feet long but nonchalantly passing before our emergency stopped vehicle, crossing the road like the king of the jungle that it now is. We came across the La Plage restaurant in Mandrem, serving perhaps the best European food in the state to the best-dressed Europeans who can be found in quite a state. And we encountered the halcyon Cape Rama Fort struck out in south Goa's isolated littoral peninsular, a rare place to stand and scrutinise the big blue Shangri La.

In the car we really found our feet. We were able to circumvent the pestering cries of the street vendors and taxi drivers and feel a little less like tourists. We were able to stock up on shopping without having to bawl at the bus driver to wait while we loaded it on and bawl again for him to remain stopped while we unloaded it off. We no longer had to suffer the white-knuckle journeys to the city at the whim of terrorising rickshaw drivers with one eye on the road and the other wildly oscillating between the hands of fate. Instead, we drove our novelty-mobile where and when the mood took us, drawing incredulous stares in our wake, attracting too much attention to feel normal, but feeling somehow special instead. True, 'special' either meant 'remarkable' or 'pass-remarkable' depending on your viewpoint, but it was special for us, and living with our At Home He Feels Like A Tourist predicament had finally started to become fun. It is perhaps a gloomy presage that to feel truly comfortable we needed a car, remaining slaves to our machines, even in such an un-mechanised country, but it couldn't be avoided.

****

In a moment of sentimental madness I had invited my father to come to stay with us. Actually I insisted he came, ordered him a flight ticket and ordered him to come. It seemed such a good idea at the time, such a good idea to invite a seventy three year old Irishman with a ferocious temper and curmudgeonly disdain for anything new. My father, a man who had never ventured further than England in his three score and ten, who does not eat anything other than boiled bacon and cabbage, who has the palest skin imaginable and is explosively opposed to sunscreen, dismissing it as 'marketing nonsense' and who, faced with the exponent of any religion which is patently not Catholic, is likely to lash out first and ask questions later. Add to this his unpredictable countenance and most predictable contrariness, and his assured divergence with any views that I myself might hold and I was beginning to wonder whether it was really such a good idea that he visit. As his arrival loomed I began to grow more anxious by the day.

The truth is I really like the old man, despite his difficult nature, and I wanted him to take this opportunity to have a look at another world, a chaotically delightful world in my book and I wanted to know how it figured in his. I felt he might be the perfect antidote to the polite holiday small talk I was getting from other elder holidaying visitors, I knew he would give me some rectilinear opposition, and he didn't let me down.

It started at the airport. Julie and I waited in anticipation for his arrival. We waited two full hours while two planes landed together in the small military airport, causing chaos. We searched eagerly for his sure to be smiling face, it did not appear. At length and after some concern we saw him. He did not look all that happy. We jumped around, waving and smiling, furiously rapping on the glass that separated us from him, he of course didn't hear us; he is, we remembered, all but deaf. Unperturbed we continued to cause a commotion and asked other passengers to tap him on the shoulder, a couple looked doubtfully at his unamused physog and looked the other way. Looking at him again I didn't blame them. After another painful twenty minutes watching him try to work out which of the two carousels his luggage was going to appear on, which was as frustrating as watching a baby trying to force a square block into a round hole while the round block rests ostentatiously on the floor beside it, he simply gave up and stood still for a couple of minutes, seemingly in a daze, unaware that we were watching, as if his mechanism had shut down and was suddenly on standby. We were unable to intervene, barred by the customs and excise counters. Then all at once he seemed to wake up, as if someone had pressed a button on his remote control, and he simply walked over to the end of the carousel and collected his bag where it had been transferred to the floor by the helpful ground staff who were presumably sick of watching it go round and round. Within minutes he was upon us. We felt relieved. He looked tired...

"Get me away from this bloody stink" he barked as we waited for Julie to collect the car. We were standing by an open drain.

"Err...you might need to get used to that smell cropping up from time to time..."

He looked alarmed. He looked even more alarmed as Julie approached in the car. He has long been an opponent of my predilection for timeless, elegant vehicles and he reaffirmed his position once again.

"Oh my God, another old banger. What is this? I can't sit in that? Is it safe? I'm amazed it actually works."

"Thanks Dad, we're really proud of it." I felt my back rising in my defence and took a deep breath, "Look, just get in, you'll enjoy it..."

The drive back to our house was an eye opener for him, no doubt about it. Indeed, it had been an eye opener for me the first time I arrived here. I let him take it in without interruption, I let him do the talking. He was most perturbed by the female road-workers, I explained that they get only 100 rupees per day while the men command a princely 150. We stopped in a roadside café to let him get a little more oriented by buying him a large whisky. His eyes lit up as we told him the price. Fortified, we continued. He began to relax a little more until we took a short cut through the fish market at Betim, the smell is all consuming and it didn't help that we were in an open top car. We finally reached home and decanted him and his belongings. He took a look around, clearly disorientated. His eyes rested upon some coconut stalks that we had collected from the beach. They were really just a bunch of sticks that hold coconuts together on the tree, we just thought they looked kind of nice and bought them home for ornaments.

"What am I looking at here, bird's eggs?"

As I looked incredulously at my father I realised what I had done. I had bought him to India. He is a 73 year-old man who does not even eat tomatoes. I became a little scared at the enormity of my task. How could he make sense of what he would see in India when he thought a pile of sticks looked like bird's eggs?

"They're sticks dad, sticks from a coconut tree. Are you feeling tired after the journey? Would you like a cup of tea? …"

The next morning he was up with the larks, and the crows and the dogs and the bread man's squeaking horn. He had made himself a cup of tea and was marvelling at the world outside the front door. This became his routine over the coming days. And to my surprise, he settled very well into the role of observer, an opinionated observer of course, and quite in keeping with his appearance as a latter-day Mr. Magoo (particularly in our little car in his Panama hat) he came out with some perfectly blinkered observations:

On Cows

(Cows are everywhere in India, considered sacred by Hindus, they feed themselves from dustbins on paper or cardboard as an available alternative to grass)

"I don't agree with them putting animals before people."

"I wouldn't eat cardboard, even if it was on a plate."

On Hinduism

(A devout Catholic himself)

"It's just ridiculous adoration to me, all this creating altars and worshipping Gods at them. Nonsense. And weird too."

On Food

(He eats hardly anything, if it's not bread and cheese he probably won't even try it)

"Well I have to say the food here isn't very good at all."

On Indian People

"I hit an Indian once, called me an Irish Cunt."

On Landscape

"This could be Woburn or anywhere like that."

"This reminds me of Ireland." (numerous times)

The truth is that my father was a real pleasure to be with. He made us look at things through very different eyes. He visited what to his generation was a third-world country without prejudice or expectation and just got on with things. He didn't moan or whinge once. He didn't get a single mosquito bite while Julie and I suffered so many as a result of his consistently leaving the door open. He didn't get sunburned despite leaving his feet in the sun (I was forewarned of this habit and kept moving him) and in spite of his repudiation of the worth of UV protection. He drank lager without moaning despite his customary polarity to any alcohol which is not brown, and most importantly he was nice to people. Yes, he was complimentary both to and about people. Most amazingly of all, he made the effort to listen and understand people, and at least to some degree, succeeded. With some effort and ear straining, he could hear and understand what people were saying. Obviously, he didn't necessarily like it...

During his two-week sojourn, we took a trip to Hampi in Karnataka. I had been wanting to visit and this seemed like the perfect opportunity to show my dad some of the countryside and try him out on an overnight stay in an area which, although full of tourists, is categorically not tourist-friendly Goa. It would also give us a rest from responsibility as we placed our trust in the hands of Peter, our confident and judicious chauffeur and unlikely tour guide. Peter had secured what he defined as a ten-seater vehicle with air conditioning from his boss to make our ten-hour drive bearable. He arrived in a Toyota Qualis which is another of those unavoidable sports utility vehicles which parents use to pick up their children on treacherous off road school runs, with a four wheel drive option for those perilous potholes in Sainsbury's car park, an Indian market version of what Viz Comic terms 'Sledgehammers to crack nuts'. The ten seats referred to comprised a front and middle bench seat with presumably room for three on each, and two tiny benches in the back facing sideways and accessible through the boot only, the type of seats that induce travel sickness within two hundred yards. A ten seater by Indian standards certainly, indeed, I have seen at least fifteen Indians crammed into one if you exclude the kids, but a maximum of seven people from anywhere else, including the driver, would be comfortable on a short journey, four if they were all Americans.

Just as we set off, Peter stopped at a Catholic shrine to light incense and pray for a safe journey, while we remained in the vehicle and raised eyebrows to one another. We knew we were in store for at least a ten hour journey, which would start by rising into the mountains on the Goa/Karnataka border and descend onto an agricultural plateau to drive unswervingly for many hours until we reached the boulder strewn landscape in which the former Deccan capital of Vijayanagar, near the village of Hampi, sits. But we had no idea of the condition of the roads and the suspension felt rather hard, a startling oversight in such a vehicle which, when we most needed authenticity, was clearly all façade; the mock Tudor house of the auto world. We climbed into the Western Ghats and became embroiled in the precarious game of overtaking the painfully slow iron ore trucks heading for the border. The loose red dirt of the roads added to the treachery as the huge wheels of the bright orange trucks in front furnished the windscreen in a veneer of orange soil each time we approached one for overtaking. While Peter's driving is principally circumspect and attentive, he is still Indian, and as such appears to occasionally leave things up to fate. As he blindly darted between the sluggish cargo laden uphill lorries and the empty, horn blaring, turn-off-the-engine-and-remove-the-keys-SHIT-the-steering-lock-has-engaged, giant death mobiles hurtling downhill towards us at terrifying speeds, we pretended we could handle it and that this was all part of the fun. Once we began our own descent, however, we began to reach some impressive speeds which I feared were above those reached in the vehicle manufacturer's brake testing sessions and I felt the need to take Peter in hand lest we fail to arrive at our destination on account of being dead.

"Peter, can you slow down please? I think I speak for all of us when I say we are nothing short of terrified."

"Is more dangerous to go slow no? Truck behind is not using engine, saving petrol, it cannot stop so much of weight with its brakes, if we go slow they will hit us no?"

"OK, point taken, just an enquiry. As you were."

I resorted to closing my eyes and thinking of sunny days, taking long deep breaths. The problem was the sunny days would quickly evolve into leafy mountain walks, and then mountain roads, where I'd turn a corner and see an out-of-control lorry hurtling towards me.

We finally reached less life threatening terrain, tilted at a far more sensible angle and as I opened my eyes I could see miles of flat farmland aglow in the resplendent yellow of sunflowers. We proceeded at a rational velocity along a thoroughfare of unrelenting straightness for four solid hours without interruption save for the occasional necessary urination brought on by the sub-zero air conditioning and unyielding bumpiness of the road. Peter was the first to crack, pulling to a halt and giving me a slightly embarrassed hand gesture to signify his intentions

(I love the way Indian men raise a half cocked little finger to imply needing to pee, like Eric Bristow throwing darts or a posh aunt drinking tea, such a modest gesture), and we repeatedly took his lead.

The drive afforded us a perfect agricultural motion picture as we bounced along, waving to the farm hands and shepherds, and overtaking the biggest moving haystacks known to humanity. Approached from behind, they emerged as leviathan Magic Roundabout Dougals, each scurrying in an ill-tempered search for sugar lumps. These haystacks were bigger than houses and were mobile by virtue of having been constructed around a lorry with a small gap thoughtfully left for the driver to peer through, and they housed whole villages of workers on top, getting a well earned rest before presumably dismantling these straw mountains upon reaching their destination. I only hoped it wasn't too far, for not only were they shedding load like moulting camels, but they were swaying and swerving so violently with the encumbering weight distribution that they were a serious hazard to both themselves and the other road users, namely us. Safety is an underrated concept in India, and the concept of 'carrying a dangerous load' that prevents us Brits from driving with an open boot or a ladder protruding through the passenger window is utterly unfamiliar in India, where anything goes, and very often, everything goes. In fairness, I remember a window cleaner in London who used to ride with ladders strapped to his bicycle, and certainly the most severe case of overloading I have seen was actually in Morocco, where I beheld a moped so laden with mint leaves that it occupied both sides of a two lane road as it progressed, and bestowed everything in its wake with a fresh minty flavour. But India is surely King, its dedication to the dangerous load would see the subcontinent win a rare Olympic Gold for sure. For the perfect example one only has to look at that famous image that did the email rounds some years ago of a donkey pitched into the air having been shackled to a far too heavily laden cart. There are of course the legendary sights of full to bursting trains with hundreds of not so lucky roof mounted hangers on and downright unlucky carriage-side mounted hangers on. Yes, London underground trains are sardine packed during rush hour but these pale into insignificance before the Mumbai commuter trains. A six seater rickshaw or Tempo will regularly take twelve, and I once counted twenty one people inside a ten seater loose topped off-road vehicle that I was following, crossing extremely rough terrain with one man, who obviously didn't want any more kids, straddling the back gate, one foot inside the vehicle, the other dangling outside, every bump administering a sharp blow to his uncomplaining bollocks. (Perhaps even more astonishingly he was passed a nauseous child at one point who he held out the back of the vehicle by the waist, de-puked, shook off the drips and passed back inside).

I love the way everyone is just so nonchalant about this too, helping more complete strangers to board a bus to add to my own mounting discomfort is way off my agenda, but commonplace here, and the way people witness impossible sights without batting an eyelid, I find even more astounding. Some of my favourite examples are: the three boys I once saw riding a pushbike, the last one carrying a huge floor-standing fan behind them which was turning in the wind looking like some ill conceived Heinz Wolff propulsion system. Or the man I saw trying to turn left into a small lane on a bicycle on which was mounted an gigantic piece of plastic pipe, some five metres long (I do not exaggerate), or my absolute favourite, the aforementioned two men and a fish tank full of fish on a moped. In this comedy double act, the pillion passenger had a three-foot rectangular fish tank on his knees. The tank was filled with water and had a lid. As I pulled alongside, I realised that it was also filled with live fish, and that the inevitable bumps in the road were causing the water to lap up the sides and slosh out of the none too secure lid. The man's legs were visibly soaked. He held the tank balanced using only his left hand, while in his right hand he carried a plastic orange flour sieve, so did not even have a solid grip on the tank - as I stared at him incredulously from my car and started to whoop with pleasure, he looked at me as if I were the outlandish spectacle and tapped his friend on the shoulder telling him to look at the funny laughing foreigner, meanwhile all other traffic around them was totally unconcerned by their marvellous spectacle.

As we trundled along, casually overtaking the Dougals and their waving parasitic passengers, I surveyed a calmness around me that I realised I had largely left behind when I left England. It hit me that while Goa is rural and peaceful, my experience of it had been predominantly as a hot and hassled tourist, rarely afforded a moment in solitude, either on the beach, in the countryside, the street or even in a bar, without being continually interrupted either on some pretext for business or out of friendly curiosity. I was again made aware that in Goa, peace was only available to foreigners when in a car or in a building, and mulling this over I fell into my first homesick reverie. Staring out of the window in silence for an hour I became quite melancholy, reflecting that in over three months I had achieved very little, I wasn't really writing much, I had taken a job that transgressed the whole intention of the trip, and here I was wistfully lamenting the loss of freedom, incarcerated inside home or car, or forever beleaguered by pests, both human and fauna. I began to consider whether I was in the right country after all.

We drew to a halt for the Nth time as another assembly of sheep-goats engulfed the car, and as I blankly stared at just another shepherdess in my downhearted daydream, she, suddenly shining in brilliant cobalt clothing, cracked me the biggest, most beatific, most genuinely friendly smile possible, and she had clearly smiled it for me. In reflex I reciprocated with a beamer of unprecedented magnitude and my sombre mood vanished in an instant.

"Did you see that woman smile?" said my father in high pitched incredulity, "That was just wonderful."

"Amazing", I replied, "everything about this country is amazing."

We ploughed on past the fields with their perfect uniform irrigation channels, past vast quadrants of red chillies drying in the sun, past further swathes of screaming sunflowers and more sheep-goats than necessary for a lifetime's counting to sleep. I call these curious creatures sheep-goats because their thoroughbred is quite ambiguous. Alive they are tall, scruffy and ill tempered like goats, but fluffy and skittish like sheep at the same time. In death they are called mutton, but as most will concede, save perhaps for a few obdurate citizens of Lucknow whose kebabs are to be protected as national treasures, they taste just like goats. There are of course many different breeds of goat in India as elsewhere, but these are all easily identified as goats, unlike these apparent hybrids that I am hesitant to label a flock and equally reluctant to call a herd, so will settle instead for *Flerd* as the most appropriate collective noun. Indeed, many people when asked (and I do make it my business to ask) say that mutton is goat meat and that these creatures are goats. When I ask what the other more recognisable goats are, they say that they are also goats, and thus, mutton too. When I begin to point out that not only is goat meat not mutton, but these two different creatures cannot possibly both be mutton and taste the same because they are not equivalent in looks, I am reminded that all meat is simply labelled 'Non-Veg' in India and has very little taste of its own anyway, hence the delicious melody of preparations available to improve it. So a *flerd* of sheep-goats they will have to remain. Just call things whatever you want, what the hell? Everyone else does.

We rolled past Londa and Alnavar, to Dharwar a typically hot and smelly conurbation, where my staunchly scrupulous father witnessed his first broad daylight cop corruption. Quite lost and employing the 'ask directions every few minutes' method of navigation, Peter unwittingly bumbled through a red light, and despite the mass of two-wheeler traffic audaciously doing the same, a cop spotted our out of state plates and pulled us over. Peter looked at me through his eyebrows and got out to see how much it would be this time. As Peter handed over a one hundred rupee note quite openly, and the cop pocketed it equally openly, grabbing my dad's seatbelt from behind and leaning heavily on the door locking button was all that I could do to restrain him from getting out to confront and redress the situation.

"Let me out, I'm not having that", he bawled as he struggled with the door.

"Dad, you're not at home now, just leave it, it will be better for Peter."

"No way, I'm not having it." He opened the window over which I had no control and screamed at the cop. "You filthy bastard, I saw that, I'll have you in court you despicable scum. I'm not having that, its extortion you filthy pig…"

Peter climbed quickly back in and started the engine in haste as the cop came towards us wearing an unholy expression and reaching for his *lathi*.

"You're all clear at the back Peter, go, go…" I barked as we screeched back a few feet before slamming into first and speeding away.

"You filthy bastard, give me your number, I'll report you for that, don't think you'll get away with..." his voice became quickly inaudible to the policeman who offered a sarcastic wave as we sped off, my dad's incensed crimson head still thrust fully out of the window shouting expletives toward deaf ears.

This episode stemmed our hunger somewhat and we shelved plans for lunch, electing instead to drive on to Gadag which Peter felt should take around 2 further hours. About an hour into the drive, my father, eventually calmed after his outburst but still shaking his head at our acceptance of what we tried to explain to him would otherwise be a harrowing ordeal where Peter would have to wait hours at the police station for his licence, if he ever saw it again at all, mentioned that he would be needing the toilet fairly soon in a sitting down way, and that while he had not suffered any sickness thus far, this morning's constitutional had been a comparatively loose one. By the time we reached Gadag, the atmosphere in the car had become rather tense as we all secretly took note of the changing facial expressions of the elderly gentleman in the front. As none of us knew the town, and none of us could read Kannada, the local language, finding a restaurant that was likely to have a suitable toilet was proving something of a strain. Finally we opted for what seemed like the best option, a large Pure Veg establishment, which I thought was unlikely to have a Western style lavatory, but might at least have an apposite squatter. I was right, and after a painful wait for it to be vacated, armed with a toilet roll that Julie had had the foresight to provide, he left us at the table. After ten minutes I became worried in a reverse paternal way I had never before experienced and knocked on the door to see all was ok. Reassured it was, I returned to the table to eat. We did not order anything for Pop, this is complex at the best of times and I assumed a packet of plain crackers might be the best fare under the circumstances. He eventually emerged smiling and relieved, washed his hands, grimaced and returned to the toilet. After a couple of similar false starts, or rather false ends, he at last joined us at the table and sat tentatively down. He smiled, put his hand on Peter's shoulder and said, "I'd appreciate it if you could drive slowly for the next few miles my good friend."

Rather than drive straight into Hampi, we had decided to stay in the only sizeable hotel in a nearby town, described by my guidebook as 'charmless Hospet'. I would have to concur with this depiction, but everywhere has appeal for someone, and for my father, Hospet was his first taste of a typical Indian town by night. Before eating dinner that night I took him out to help me find some cigarettes and some Imodium in case tomorrow brought further evacuations, and as we stood in the street watching the crowded little medieval bazaar buzzing in the darkness, illuminated by the shadeless hanging bulbs of the stalls, I could tell that my dad was in absolute wonderment. As he saw *bel puri* vendors and cycle rickshaw wallahs for the first time, the puncture repair man, the street cobblers and barbers, the *paan* seller and the hectic little pharmacy, so many vendors and so many menders, so many people and so many animals, he turned to me and said, "I can't believe I am actually standing here watching this with my own eyes, it is so different from how it all looks on TV, so much more real and far less horrible, I feel like my eyes have been opened and I will never forget this." From a man not given to emotive outbursts I knew this was sincere, and I knew I had been right to risk bringing him here. For me, the commonplace bustle of a typical Indian street is more captivating than any in the world, and so it seems, it was for my father.

Hampi is a village with the privilege of being central to the ancient ruins of Vijayanagar, the erstwhile Hindu empire that grew from its 14th Century beginnings to control the bulk of southern India during the 16th Century until 1565 when the brutal administration was curtailed after a bloody battle against the conquering northern Muslim coalition, who promptly moved in to the largely trashed city, built a few incongruous buildings on top of some existing ones and finally abandoned the whole thing to some locals chiefs who, unable to restore the city to its former greatness, slowly watched it deteriorate and die. (Apologies to any historians for my simplistic and quite probably inaccurate portrayal of this grand Hindu empire but, you know, there are history books for that sort of thing).

Today Vijayanagar or Hampi is for me the most magnificent lost city imaginable. The Vittala temple with its resonating musical pillars, the Hanuman temple on the mountain top, the Lotus Mahal and Elephant Stables, the Queen's Bath, and the enormous army barracks with its numerous covered marching boulevards and other vistas of pomp and circumstance are all located among the most mysterious topography of rocks and boulders conceivable. Despite the deficiency of foliage in this harsh stony place, the ruins somehow conjure Disney's 'I'm the king of the swingers' interpretation of Kipling's Indian folk tale, and swinging with monkeys the place certainly is, the elaborate skyward pointing roof of the Virupaksha Temple is a blur of Langurs and Macaques in spiraled pursuit of each other, but while they occasionally come down to take offerings of food from pilgrims, they don't hang around and perform or say thanks, they simply grab the banana and off. Here I'm reminded of my friend Anupam who so wisely pointed out that "Monkey can touch to you, but you not touch to monkey, ha ha, no way, you not touch to monkey." Quite.

Being inland from the coastal hue, the sky here is a deep clear blue, the like of which turns any old happy snapper into a photographic professional, and the clarity of light transforms the human retina such as to firmly put in place the old rose tinted spectacles. This of course means that whatever your mood, Hampi will enhance it, and it is noticeable, really, everyone here is visibly happy. Walking up the main bazaar we encountered scores of backpackers and pilgrims (Hampi is still a very significant Hindu pilgrimage site) happily smiling into the sun and looking for the whole world, generally very content. And it is contagious, despite the inevitable hassle by a multitude of vendors, we ambled comfortably through and into a small restaurant to escape the heat and reinforce our constitutions. As we sat down I opened the menu with relish. My excitement was not in expectation of good food, although I was sure that it would be, but was in anticipation of embarking once again on my much favoured hobby to see which uproarious misspellings this particular carte du jour might yield. I concede it is a cheap gag to be continually amused by mistranslations and phonetic misspelling, but I have made it something of a hobby. I am a simple pedant, who, if offered the opportunity, would still deign to watch 'That's Life'.

Moreover, I would keenly urge anyone to resort to being just a trifle more childish when confronted with such bill of fare blunders in future, don't ignore them, enjoy them, even the most simple examples can brighten up the darkest of days. For example, the menu in question threw up but one incongruity, 'chicken toast', not funny you might think, but what can it actually be? Really, chicken toast? What is that? It could so easily have been overlooked and squandered, such a waste. Scour away hair-splitters, I salute your sophistry with a light heart.

My favourite examples range from the basic vowel trade of 'Hat Dog' noted in a Turkish hotel, so disarmingly simple, to the bewilderingly bizarre 'Whole fish in the shape of a squirrel' that had been translated from the Cyrillic on the front door menu of a (sadly) closed restaurant in Sozopol, Bulgaria, to the brutally honest, 'garlic fried calamari (squits)' in San Sebastian. But again, India tops the pile for sheer volume of opportunity. The fact that there are eighteen official languages, many with different alphabets that need to be translated phonetically to be read by non-speakers, dictates that there can be no commonly applied rules over spelling anyway. For example, there is no agreed manner in which to spell the term *bhoona* or *bhuna* or *bewna* or *boona* or *buna* of *booner*, (a widespread cooking style which entails part broiling and part frying of meat in thick gravy,) so there is really no hope at nailing an agreed credo for anything else. Add to this the fact that many restaurateurs do not use English as a first language and you have yourself some real prospects.

Rather than list a hundred or so pernickety and often quite pitiable examples, I have chosen instead to publish part of the extraordinary tariff as offered at the Marita Beach Shack in Palolem, Goa, which, I'm sure you'll agree, really does justice to this phenomenon:

Continental Breakfast

Choice of Juices
Choice of Egg Juices
Carrots
Toast
Salad

American Breakfast

Choice of Juices
Choice of Egg Juices
Circles
Toast
Salad

Lunch

Backed Beam & Chip
Mass Wine Chicken Curry
Sore with Boiled Veg
Thai Curry (served with S Stem rice & Noodles Got One)
Creamy Backmol Sauce Top with Cheese & Bread Crab

In the less crazy cafeteria in Hampi, the 'chicken toast' turned out to be a quite palatable toasted chicken sandwich which our finicky elder even managed to consume. Nutritionally equipped, we set off for a first-rate few hours of extreme sight seeing, excessive photography and overall wonderment before heading back to Hospet, and the following morning, all the way back to Goa. At the hotel, Peter had got wind of a different route home through the teak forest and towards the coast at Karwar, a slightly longer route but with Forestry Commission quality roads, we made it back in just over eight hours, exhausted but at ease.

Peter had to go to work the next day, and so, I remembered, did I.

We drove my father to the airport the next day with a mixed sentiment of relief and sadness from both sides. The effort had been worth it but we were all agreed that two weeks was sufficient, any longer might have become 'tiring' as my dad graciously put it. An hour after his plane left, I myself flew off for a two-day stint in Pune to meet clients and generally sit around and nod seriously as we all put forward our most earnest intentions. That night I told Sarasvat how I was half sad and half glad to see the back of my dad after only two weeks and quickly wished I hadn't as he informed me that in a week's time both his parents and grandparents were coming to stay with him for two MONTHS.

In a stroke of wonderful good luck, the clients cancelled at the last minute, having not secured travel visas in time (nonsense, you just go to the embassy and get them on the day, a ruse if ever I've heard one), so I stayed a couple of days getting to know everyone at the office better and flew back to Goa in good time for Christmas and with no further work trips planned until mid January. This was fast becoming the best job I had ever had.

****

Earlier in November the Indian professional football season had started and while I must flagrantly state that I am not an advocate of yob culture, I am a firm supporter of football. I have never endorsed the hooligan constituent, and I am very sceptical about the patriotic jingoism surrounding the national game, but I love the game itself and I love *going* to football. As a supporter of Luton Town, the glamour of the big game is often elusive. It is perhaps this that has spawned my fervour for watching games in foreign countries. I have seen both professional and amateur games in many European countries and in Africa, the United States and Asia. The level of the game is largely unimportant for me; it is the difference in atmosphere that I seek. Hungarian professional football might not be dissimilar to Polish, but the ambiance of the crowd could not be more distinct. The San Siro is a far cry from Sunday league in Como, but the zeal is almost identical. Beach football in Morocco is in stark contrast to beach football in Brazil, which in turn is far removed from beach football in Wales. Football in India is again quite unique in both style and character.

Cricket is quite obviously the national game of India, but football is on the up. The English Premier League is watched by millions on TV and football shirts are everywhere. Bengal is the undisputed capital of Indian football. The two biggest teams in Calcutta, East Bengal and Mohun Bagan, are regularly watched there by up to 120,000 fans. Kerala is also a footballing state, with many leagues and a history of top national players. I saw my first Indian game in Kerala some years ago, a painfully poor performance by both teams in forty degree afternoon sun, watched from a single crumbling terrace, stained red with *paan* spit. The crowd of around 100 were enthusiastic and in good voice but were hopelessly drowned out by the relentless, thunderously amplified devotional chanting from the neighbouring Hindu temple.

Goa is also very much a footballing state. Of the fourteen teams in the National Football League, the only professional league in India, six are from Goa. The Nehru Stadium at Fatorda is a 15,000-seat arena that regularly sees crowds far exceeding that figure (safety is always second to sales in India) which is shared as a home ground by all six teams. We decided to go to a game and as our friend Anupam is an avid supporter of his home town Calcutta's East Bengal we picked him up and set off. Driving to a football match is always a thrilling mix of expectation and trepidation, but driving through thousands of Indian supporters in an ancient convertible clunker is really quite another experience altogether, needless to say we caused something of a stir, attracting more attention than visiting fans should, so with our 'best seat in the house' tickets, patent white skin and token Indian mascot looking something like penny-for-the-guy in his outsize West Brom shirt, wading through the baying cheap seat queues felt a little unnerving, not quite disembarking at Bermondsey Station and tiptoeing towards the New Den perhaps, but a bona fide football away-day experience nevertheless.

Inside the ground things were somewhat different, the forty rupee outlay secured us a shaded pitch view and a fixed plastic seat as opposed to eye level sun penetration and a scorched arse from burning concrete, and a surfeit of chanting 'water, water, water, water, water' and 'smosa, smosa, smosa, smosa, smosa' boy vendors. The support largely comprised the educated set, the nouveau riche, a smattering of bloated tourists with more than a few Oafs among their number, and a diminutive, tight knit, insane fraternity with enormous home made flags, air horns and a vivaciously animated comportment to rival Ren and Stimpy. The assembly was, of course, predominantly clad in Manchester United shirts.

Today's home team were Dempo, a team sponsored to the point of total ownership by the Dempo Group, Goa's iron ore and media giant. A team with two Africans and one South American among their rank who had recently been able to bolster their side due to 'unknown investors' (tax free cash from the company according to my neighbour) and had high hopes for the season. East Bengal were and are an altogether more accomplished set, with India's preeminent home grown player and international star, Baichung Bhutia among them. Bhutia played in the English league for some time, getting the occasional run out for Bury, which is probably the best yardstick for measuring the class of football on offer.

The game turned out to be a ding-dong 3:2 thriller going unexpectedly in the favour of Dempo who somehow managed to escape a thirty minute East Bengal onslaught comprising no less that twelve clear cut scoring chances. Given the heat, humidity and intensity of the game, and add to this the overarching lack of skill and fitness, and it will come as no surprise that after seventy or so minutes, all substitutes had been used and there were at least three players down with cramp at any one time. Despite a real opportunity for East Bengal to capitalise, using their mildly superior skill and experience, the final five minutes was played literally at walking pace, something I have never witnessed before or since, and such sauntered tackling really is a spectacle to add to life's 'do not miss' list. In consequence of their spirited win, Dempo became our adopted side for the coming season, a wise choice as they finished second to East Bengal in the league and showed a determination often very much lacking among other competitors.

A definitive Indian style of play is not at all easy to decipher. On the beach I would say it is leisurely nonchalant with occasional bursts of chaotic effort, and nobody watches, everyone joins in. In the parks and dusty pitches, it is hotly contested, physical and boisterous, but with a reticence and rationality which demonstrates a sensibility quite different to its European equivalent. The recognition that opposition players will need to keep their limbs intact for use at work the next day, a sensibility that the spectators themselves endorse, the goading and chanting being very much in the same spirit. The professional game however, rarely exhibits the skill and strategy to illuminate the crowd to full beam. The potential spectacle is immense, but rarely delivers. We attended seven or eight games in the earlier part of the season, after which, to be quite honest, the novelty wore off a little. Even a jaunt to see India play Singapore in a world cup qualifier in a thronging, twice sold out stadium, threw up some very forgettable football, and an atmosphere which steadily lost momentum as the crowd became slowly more interested in talking about cricket.

Professional Indian football has a way to go to really get my vote. It appears to be played in a real 1980's 'hit and hope' fashion, the long ball played route one and a defensive error being the predominant scoring technique, which is simply not very interesting to watch and unlikely to kick up much of an atmosphere. Amateur football played in the local parks in Goa is a much more exciting spectacle, holding, as does all localised sport, the great possibility of getting one over on one's friends and neighbours. But it is just so hot sitting out in the sun watching it, and so much hassle having to continuously answer questions about how many times you have been to Goa and fend off continuous appeals for cigarettes and rupees. For the time being, I'm sticking to cricket.

And cricket is very easy to stick to. It is everywhere. Absolutely everyone in India is into cricket, the game is inescapable. From the *maidan* to the beach, cricket is played on every available flat surface by kids and adults alike. Every street is a potential track and every wall is a potential wicket. I have even seen traffic patiently skirting around an impromptu knockabout on Marine Drive, Mumbai's busiest road. It is endorsed by all, not least the national team, whose players endorse every product available on the Indian market. Indeed, Sachin Tendulkar is a veritable publicity prostitute, appearing on roughly every third advert on television.

While I have always thoroughly enjoyed cricket in England, I have somehow never really had enough time for it. The feverish pace of the working life is not geared to leisurely afternoons watching men in white and listening to leather on willow. Even the limited overs game takes all day, and the newborn twenty-twenty contest can be said to be thrilling but at least for the moment, is just not cricket. To enjoy cricket fully one needs to watch each and every over and witness the progression of the pitch and the player's response to it, highlights have never really done it for me and it is as much as I have been able to do to loosely follow the form in the newspapers. But here, here I had the time, and where better than India? The land where the names Boycott, Botham and Border are as well known as Gupte, Gavaskar, and Ganguly. A land with two TV channels devoted to the sport and a further three with daily coverage, not to mention the abundant news specials, quiz shows and talk shows dealing exclusively with cricket. A land where everyone is an expert, and the question of who should open the batting with Dravid is debated daily in eighteen different languages and hundreds of different dialects.

So we sat back to enjoy. The first test series to appear on the telly was England's tour of Sri Lanka, and despite the TV station broadcasting adverts after every over at deafening volume (at last I realised what the mute button was for), I watched practically every ball. Julie too became embroiled in the drama of the game, to my chagrin becoming an avid Sri Lanka fan, lured by the impossible match winning brilliance of Sanath Jayasuriya and the steely, needling wind-ups of Kumar Sangakarra, whose low pitched cries of 'bowling' as each spun ball is collected have become the generic for 'well done' in our household.

Slowly cricket replaced football as our interest grew. We stopped going to the sports bar on Saturday nights to sit among our swollen, scarlet, stewed compatriots to watch them bawl at the screen and brawl with each other, instead opting to stay home and watch civilised repeats of past test glories on Cricket Gold. In fact we gradually began to shun the whole tourist debacle as the sham we knew it to be, eventually even eschewing the beach and its pleasures, spending more and more time kicking around the house, lazing supine on the sofa, forgetting about our past lives and even our current coveted pursuits, our indolence inexorably aided and abetted by the Gentlemanly conduct of cricket. And long may it continue. Long, long may it continue.

Chapter Eight

Christmas is a comparatively low-key event in India. Well, low-key when compared to Europe. With an estimated eighty-five percent Hindu population this should not come as a surprise. But Goa is again different. The English Imperial rulers never saw themselves as missionaries; at least they were not on a religious mission. Their strategy was to allow their captives to coexist without cohabiting, and religion was a sure fire way to segregate while maintaining respect. There is no doubt Christianity was already prevalent within India prior to colonisation, particularly in the south, and there are around twenty million faithful today, but the British were not responsible for its active proliferation. It seems the Portuguese settlers of Goa had no such strategy, when Vasco Da Gama first set foot on Indian soil, he did so at least partly in the name of his religion, and it is Catholicism that dominates Goa today, both architecturally and spiritually.

So Goa has a very busy Christmas by Indian standards, but it also has a rather swinging Diwali. This festival of lights is what a couple of my Hindu friends rather crassly call 'our Christmas', and I think that by this they mean their biggest annual religious celebration. Hinduism is of course a fantastically complicated religion with over 300 million deities. And with nearly 800 million staunch worshippers worldwide that's more than One between three, a pretty healthy ratio compared to One between something like two billion for Christians. But that's a lot of birthdays, and events to celebrate, and unsurprisingly, Hindus from different areas celebrate Diwali and the many other festivals in very different ways. For Goa's Hindus, Diwali is both initiated and chiefly celebrated on Narak Chaturdashi, a night when huge, gaudy, firecracker-stuffed *papier mache* effigies of the demon Narakasur are paraded along the roads accompanied by deafening music and finally burnt or cremated at dawn.

This was something I had to see, so on a semi-deserted street at 11pm when all the holidaymakers were securely back to their hotels, I hailed a lone taxi and set off across the tracks to the Hindu quarter, or in real terms, the Candolim to Panjim road, which to my surprise, was already chock a block with tourist taxis out to see the show, just like me - it seemed the package holiday companies were in on the deal. The route was alive with immense garish reproduction fiends rigged up on terrifying scaffolds, invigilated by deeply excited kids with an indomitable intention to get a few rupees out of the *farangs*. I began by eagerly stopping at each one and congratulating its builders, giving a few coins out in encouragement until the taxi driver revealed the enormity of the spectacle to me from the top of the hill - there must have been three solid miles of demonic pageant, a phantasmagoric pandemonium of pyrotechnics, and the champions were clearly some way ahead. Becoming more selective, I still could not resist stopping to see every second or third one, such grotesques being so acutely uncommon in my experience. At Nerul there was a bottleneck and my taxi queued for some twenty minutes before the source was finally revealed, a twenty foot high devil, glossed in lurid purple with flashing red eyes and doing a very vigorous and very hazardous pelvis thrust with the aid of a complicated system of pulleys and strings and some rhythmic encouragement from the Burst Eardrums School of Music. I noted one very tiny infant whose terror had inflated a set of lungs that might very soon be enrolling at the Burst Lungs School of Screeching and determined to move on lest any of my own organs start getting any ideas.

At Betim, after some thirty or so additional *diablos*, many gyrating and vibrating with the most obscene mannerisms and in the most treacherous Petrol Generator + Naked Electricity ÷ Papier Mache & Straw Stuffed with Fireworks + Too Many People Smoking = Don't Hang Around equation, there lay in front of me the *piéce de résistance* of the mock spirit world; the superlative ersatz effigy. It was at least fifty feet high with yellow check trousers and matching scarf over a red top, its head a bulging blue, and it was lurching backward and forward like some enormous choking Rupert Bear wielding a great cutlass and emitting deafening shrieks to scare the shit out of kids and tourists. Truly fantastic and truly dangerous. This is surely what religion should be like, demonstrating the principles of good and evil while simultaneously scaring the excrement out of the young and uninitiated with huge treacherous displays of terror while the faithful look wisely on enjoying their belonging. A far cry from the Easter Vigil or Pentecost.

It's a shame Hinduism is not proselytising, I might consider signing up.

I was exhausted by the whole event, wonderful and frightening as it was, but before returning home I stopped in a scruffy bar in Betim for a quick drink to wish a few local drunks a happy Diwali. As always in non-tourist bars I was inundated with offers to join people at their tables and took up the offer of the drunkest looking man in the house. Raising my glass to his.

"Happy Diwali." I offered.

"Ah yes, we must not drink, and we must drink, and must not Diwali… hallo I am Christian. No I am Diwali man, Christian man, Diwali man, boy man not. Hallo I am Christian again." He replied, chinking my glass.

"And I am Mick." I added hopefully.

Sometimes this is such a strange place.

So Christmas was on its way, and a couple of weeks later we replaced our *akash kandeels* on the porch with a star shaped Christmas lantern. Julie insisted on making ours herself from recycled rubbish, rather than buying one the same as everyone else's, with the inevitable effect of affecting unwanted inquiries along the lines of:

"Did your children make your lantern?"

"What happened to your lantern?" or

"Oh, I have a spare lantern in the house, would you like it?"

I must disclose that I am not always the world's biggest Yuletide advocate and am prone to opting out where possible, but Julie has a great love for traditional customs so I engage as best I can for her sake; I give and receive gifts, eat and drink well while visiting the family and wear a paper hat at the dinner table. In short I enjoy the family aspect but can do without the rigmarole. I was unsure as to what to expect in Goa, but I had a pretty good idea it would differ significantly from the norm as far as I was concerned, and in the interest of goodwill to all men I decided to try to take whatever it threw at me without getting cross. I felt I might fail in this respect because however hard I try, I eventually get royally fed up with the pleasantries inherent in interactions at this time and start to drink heavily or in extreme circumstances simply run away. Avoiding such liaisons would be very much preferable, but in an effort to engage with my new society, I felt I must become less humbug, and there is nowhere better to begin than Christmas.

We bought fairy lights and rigged them up around the exterior of the house along with various supplementary homemade stars, angels and the like. We procured a tiny fir tree from a garden centre that had sold out of trees of significance, draped its microscopic branches with more garish adornments and presented it on the verandah. And thus festooned, our home was ready for Christmas. What I had not figured was that displaying such concurrence with the season would of course attract like minded others wishing to share in the jollity and fun. The first evening after decoration brought carol singers. I joined Julie at the door and we smiled big plastic grins while they sang SIX Christmas carols to us before holding out their hands for a reward, upon receiving payment they started up again, and to my horror seemed likely to be repeating the performance over again in its entirety. I interrupted, "Thank you, thank you, good night" and closed the door. They shuffled off in silence.

An hour or so later and my fearfully attentive ears picked up the sound of further carols down the street, "Quick, turn the lights off and lie on the floor" I barked at Julie who was already on route to the switch as I shouted. We lay in darkness while they sang four carols to each of three neighbours before moving on. This was five days before Christmas and I began to fear that staying home in the evening would be an uninterrupted succession of devotional hawking. I spoke to Averardo my neighbour the next day to ask how long the carol singing could be expected to last, "Every night until Christmas" he confirmed and I made a positive mental note to conduct the next four nights very much not at home. I also asked him what else we could expect and he told me that on Christmas day we could expect a visit from all friends and neighbours bringing sweets and cakes. Julie and I decided we would beat them to it and visit them before they visited us on the basis that we could thus control the length of the visit, and we went off into town to buy a bagful of plum cakes as gifts.

We got wind of a very different alternative to spending Christmas in the community, which was to avail ourselves of a Tourist Christmas.

So in the spirit of investigation, and to be sure that we weren't missing out on anything, we signed up for a "Pre Christmas Bash" with our noble restaurateur hosts Bom Sporto. We arrived at 8pm on the 23rd December to find the restaurant replete with British over 60's, arranged in fours, spark in crepe paper crowns and pulling crackers like over excited cartoon characters. As we sat at a table for two (I spotted it and was indignant despite the deploring Alexander, our festive host decked out in full Santa Claus outfit), Chris and Shirley from Solihull took the stage and started Rocking Around the Christmas Tree in that good old fashioned way just as Mel Smith and Kim Wilde had done so tastefully some years before them. I raised my eyebrows, Karaoke before dinner no less.

As the first course of 'shrimp cocktail' was served, two rogue Santas attempted to gain entry to the party in an effort to court money from the guests but were muscled out by bouncers as Alexander, our own Santa, confided in me that "Those bastards always try to get in here at Christmas, why should people give them money? They do nothing apart from dress as Santa. Far too much of Santas in this town." I nodded sheepishly into my food, worried lest there should be any Santa Agro.

Between courses we were treated to carols belted out by son Edwin, now dressed as a snowman, an unlikely thing in Goa I remarked to him. "It's for the tourists," he barked at me as if I were unintelligent in the extreme. He was clearly stressed by his contradictory aspect and not in his usual sarcastic disposition.

As the turkey and sprouts were served, in the traditional boiled to death manner, it hit me that this was an authentic British Christmas in every respect, and that for those who pined for such a thing, a real effort had been made to provide it. And it would be again on Christmas Eve and on Christmas Day itself. I have to admit I was tempted to sign up for the Great Day, but just as the nostalgia and alcohol began to slurry my senses, the normally quite errant Teresa from Kent who had impressed her bodily largeness upon me on numerous occasions in the bar, struck up a most horrifyingly sincere rendition of Mistletoe and Wine to bring me back to my senses. We would certainly be spending Christmas Day at home.

We began visiting all our friends, neighbours and acquaintances on

Christmas Eve and finished off on Christmas Morning, a most enjoyable voyage of giving and receiving. In the main we had been given horrible sweets of the uncooked, sugar and water variety *á lá* peppermint creams, which were easily offloaded on the local kids. But we had also accrued a bag of cashew nuts, two bottles of beer and a plastic flower on a stand with flashing lights and the ability to play We Wish You a Merry Christmas if only Julie would let me press the button one more time (I love such things which are the embodiment of Indian kitsch and the fancy of Indian taxi drivers worldwide, by far the pre-eminent example of which was the bunch of flashing plastic grapes which played 'Land of Hope and Glory' to me in a Mumbai taxi, I am still trying to fathom the connection between the grapes and the song).

Concluding our bountiful journey at Averardo's place, we sat down to a couple of beers and a long chat about Midnight Mass the preceding night. Averardo had seen us at mass and was eager to gauge our reaction to it.

We arrived at Candolim Church in our silly car at around 11.30 to find very little parking opportunity. I wedged the car between two parked taxis, so close that I would almost certainly require a tin opener to extricate it, and as the driver of one of the taxis came running out of the church to examine the expected damage I offered a desultory "Happy Christmas" and did not look back. Candolim church is a huge one, but I had my doubts that it could hold the occupants of so many cars. As we ascended the front steps and were ushered through the empty church into a courtyard I realised that the mass would be open air, would be attended by over two thousand people, and had started some half hour or more earlier - midnight mass in Goa apparently starts at eleven pm.

I was surprised at being charged an entry fee of 50 rupees but acknowledged the sense in so doing given the opportunity. Upon being informed that the entry fee included a strip of five *housey housey* (bingo) tickets I considered it a veritable bargain. According to the ticket the bingo started 'around 2am, after mass', how long would the mass last then? Midnight mass is usually around an hour if you join after the pre-ceremony carol service, an hour and a half at most, but at forty-five minutes in and barely past the readings, this mass looked set to last hours. We were ushered to plastic seats that were filling fast and I was glad we hadn't been drinking because jammed between two families with very elderly members at each end of the row, as we were, toilet facilities had been placed beyond the realm of prospect.

The service seemed to alternate between Konkani and English, with the local language taking the lion's share while English seemed to be used almost randomly - this is another mystery of Indian language, how do people know which words to use when and in which language? The way people flit between Hindi and English is particularly amazing, and whenever I ask someone why they chose to use an English phrase or sentence among the Hindi, they usually reply 'I could only think of the words in English so I used English'. When I retort that their listener may not have understood the English, they always reply 'Everyone understands English'. I find this amazing because no one ever seems to understand me.

Anyway, the dual language service began to drag on as only Catholic services do, and the attendees' attentions began to waver as they do the world over. By twelve thirty, the old people were snoring. The old lady in the seat immediately in front of me was shamelessly indulging in some full volume pig-wheezing which kept the whole of her family in a state of permanent sleeve biting giggles, while the elder stateswoman of the family to my right joined her with some straightforward log sawing much to the amusement of pretty much everyone in the vicinity. It is great to see the younger members of the family giving the elders such respect but it is much more fun to see them laughing at them openly. As the mass progressed, more and more delegates succumbed to the somnolent atmosphere and the congregation became a caterwauling cacophony that was only gaining in volume before a sensible priest introduced a rousing hymn, waking everyone up. For the next twenty minutes we watched the old and young variously nodding off and waking with a start as the priest varied his pitch in an attempt to keep everyone conscious.

After nearly an hour of striving to be respectful and adult and follow the mass, and trying desperately not to laugh at the sleeping flock, I started to become bored and restless myself. I noted the intermittent sound of reversing cars as the taxi drivers rearranged themselves in the road. The two teenage brothers in new suits to my left had become jaded with their silent text messaging and had slumped back in their chairs to audaciously sleep without a care. I began to wonder how I could endure a further hour or more of this unintelligible and hypnotic service, but just as I had resigned to the inevitable I was saved, suddenly an intermission was announced and with lightning reactions I grabbed Julie and we made a break for it. Feeling a little embarrassed I glanced behind me as we fled to see if we had been rumbled and was terrified to see fifty or more people following behind us. Were they after us? Was this a lynch mob? We began to quicken our steps; heading straight for the car, which I was glad to see had been disentangled from its captors and was free to drive away. As we reached the car we realised that many other people had too done a bunk at the earliest opportunity, it seemed the lure of *housey housey* was none too powerful.

Upon leaving Averardo's place we would have to attempt to drunkenly cook a roast Christmas dinner on two gas rings in a kitchen with no fan and a room temperature of some thirty-five degrees. Before we braved the inevitable, Averardo insisted on showing me his collection of laminated table mats displaying scantily clad women from the 1970's, nodding and winking at me, saying "Huh? Huh?" as I flipped through and eyed each one. "Yes, very nice" I said handing them back, "very festive." "Yes, they only come out at Christmas if you know what I mean" he rather puzzling said and I passed over the opportunity to enquire further.

After a disastrous dinner that we could not be bothered to eat, we got among the drink we had stockpiled and sat on the verandah for a night of phoning family and friends on the mobile phone I had been provided with to help me conduct my new found work. Christmas had roused mixed emotions within me, I was glad to avoid the zealous commercial spirit of Christmas in Britain, and tickled that the rituals here were played out in much the same way, even Santa played his inimitable part, but I missed the family as I always do and I missed the ludicrous Queen's speech and I missed the cold cosiness of Christmas at home. I must have come to expect more from Christmas than I had let on to myself, and my blasé attitude to it had probably ruined it a little. But it was heartening to know that some 4500 miles away in a land so disparate and different to my own, the sentiment was at least exactly the same - get what you can and get it down your neck.

Much the same sentiment can be found at New Year. The New Year is Goa's busiest spell, when between December 30th and January 2nd some 2.5 million visitors converge on its hotels and beaches to pay vastly inflated prices for over subscribed and understaffed facilities. Good. I have never been able to afford to stay in Goa for the Christmas/New Year period as flights double in price and accommodation trebles and I was anticipating certain chaos. I was not disappointed.

As Christmas had ostensibly passed without incident, Julie and I decided that we would mark New Year with a great deal more effort. This is something of a departure for me because I am a firm subscriber to the notion that New Year's Eve is very much 'amateur night' where pre-paid entry to my local pub is given to part-time revellers who fill every available inch of space, get inexpertly drunk and hopelessly clog up the wheels of an otherwise well oiled night out by puking and fighting and kissing everyone and anyone, while us conceited professionals look on in horror. Where possible I now stay in with friends. But what the hell? God knows we needed some action.

Having made a sum total of zero friends in Goa who would be appropriate for such a night out, we agreed to go it alone, wear our least unfriendly clothes and expressions and try to have a good time with everyone else. There were numerous nightclubs available to us, but each time we enquired we became more and more depressed. We are over thirty years old, and the notion of paying huge sums of money to be rammed into the likes of Tito's, Club Cubana or any other "Everybody scream if you're having a blast" - "AAAAggh" club, along with hundreds of fat, tanned holiday reps, Chris and Shirley from Solihull and of course Teresa from Kent, induced immediate nausea. There would have to be some other way. Julie reminded me that there was always Vagator where there would doubtless be hundreds of young Indian boys frolicking and frottaging up against the few remaining Hippies to the sound of slamming beats and overloaded beeps. Oh, please, please, no.

My phone rang and we were saved. My workmate Manohar informed me that he had decided to come down to Goa to visit a friend who would be working on New Years' Eve, and would we mind if he tagged along with us? I kissed the phone as my answer. This meant that we would not be able to go to any of the nightclubs as they will not accept men or *stags* without a female partner on such nights, and also meant that as his hosts, we would have to show him the town. Our destiny and his - a pub-crawl the whole of Calangute, no alternative, no pulling out and no more indecision.

In preparation for the big night out I popped into Bom Sporto to wish all prosperity for the coming year and explain why we would not be spending the night there. I used my 'amateur night' theory to great effect, particularly with Anupam who seemed to understand, although one never can tell.

We met Manohar at the Capricorn Bar at the top of the Calangute beach road and commenced the swallowing. I introduced him to the superior Goan brew, Kings Beer, with its avowed German hops and reduced glycerin. Our twenty two year old Mister Ingredients marvelled at its taste and reported fully on the Indian made foreign liquor trade or IMFL as he referred to it. We had hoped to mark each hostelry with a different brand of beer until we either ran out of beers and had to repeat, or ran out of hostelries and had to retreat. But after a lengthy beer conversation lasting some six or seven bottles at the first bar, we had drank all brands that were available and returned to the Kings, where we stayed for the rest of the night. We stuck with it at my behest, because in truth, the contemporary beer offering on the subcontinent, which is all lager, and comprises mainly Fosters and Castle, (brewed in the country under licence) and the ever present Kingfisher, (whose owner the flamboyant Vijay Mallya's private yacht which he apparently bought from Richard Burton was moored a few hundred metres up the coast and in the throes of a party to which we would not be invited) is a very headachy affair on account of the high glycerin content. It is said that it is possible to remove the glycerin by submerging the open bottle in a bowl of water and watching it casually exit through the bottle neck, but the two occasions I have tried this have resulted in me holding a near empty bottle looking at a bowl full of diluted beer. So we stuck to the Kings in the vain hope of being unsullied on NYD.

We took in two further bars before reaching the first beach shack, by which time we had developed a relaxed overall attitude and if I were not mistaken, I was in fact beginning to acquire an interest in the unlikely activity of dancing. I had to admit it, I was very much enjoying New Years Eve, the revellers were certainly out in force, but they were happy and enjoying themselves, they were eating and drinking and smiling and not puking.

On the beach, the first few shacks were rammed full of overnight Indians who had come down to Goa on coaches from Mumbai, Bangalore, etc, and were staying out all night and boarding the bus home the next day. With no possibility of locating a lodge of any kind, the best policy was clearly to imbibe with great intention until the damp of the night bothered them no longer, either that, or run up and down shouting and screaming which seemed an equally popular pursuit. The music being offered in these shacks was of the fun/party/novelty/festive variety, so we edged along the packed beach to find one to suit.

Some four or five shacks along we unearthed a most pleasant surprise, a quiet shack with few inhabitants that appeared to be playing some sort of jazz music. As we approached it became clear that we were indeed correct, that is until the music stopped and nothing came in its wake. So we sat and chatted further sipping beer with poise while pandemonium raged around us on the beach. Those gathered were clearly from the amateur stable, the running around and shouting and the lack of female company made this quite evident, and although the puking had not yet started, it was surely imminent. But I didn't care, I wasn't in my local pub, I was on a beach with friends and mad people, and this suited me very well.

Calling for another drink, I asked the shack owner why he wasn't playing music to attract the omnipresent Indian crowd on a night where he could clearly make a killing, and his reply surprised me. It seemed he had no idea what sort of thing they might like to hear and appeared to have no idea how to attract a crowd which was standing only metres away (Well, running up and down only metres away) and ready to dance to pretty much anything at all in the interests of having a good time. I suggested he might play some *filmi* hits and maybe some Punjabi dance tracks, assuming that he had some, as this was certainly what I was looking for to satiate the increasingly frequent notion I was getting that dancing was impending. He led me to one of his staff who ushered me towards the clearly well equipped DJ stand, behind which stood a very serious looking chap with a skull and crossbones T-shirt, earrings, headscarf and all the trappings of a stern Indian DJ.

"Not playing any music tonight then?" I enquired.

"Just thinking about what to play, I need a change of tack, the jazzy stuff is not going down too well."

"No, perhaps people might want something a little more uptempo on a night like tonight? Some *filmi* hits perhaps? Or some bhangra of some description?"

"Do you think? I have some Punjabi MC, what do you think?"

"Definitely." I sat back down, glad to be of service.

As the first three bars of the Night Rider theme (used in the Punjabi MC track) chimed in it was a stampede, a pitch invasion, the shack filled in seconds with boisterous shaking men, arms aloft, repeatedly indicating the whereabouts of the ceiling with pointed fingers. As we suddenly found ourselves in the middle of the most popular dance floor in town, the smiling owner suggested we move to the side by setting us up a table including three free drinks. Really, the things one has to do to get a free drink in this town. As the DJ took his lead from the crowd, the pack swelled and within minutes the shack was heaving with some hundred or more pointing, patting and gesticulating dancers. The waiters could not keep up with the orders, and I could not keep up with my jigging legs.

Someone asked Julie to dance, and in defiance of Manohar's admonition and wild eyed reprove, she accepted. I watched with amusement as the excitement was turned up a notch. Drunken dancing in public seemed a fairly new experience to many participants, and having a white woman to dance alongside was a whole new experience. Manohar was unable to contain his apprehension and reproached me in sober tones, "Mick, these are people from the country, they do not know how to behave, Julie should sit down. These people are not from Delhi."

"Manohar, Julie is a fully grown woman and she is well able to handle herself with a few animated men, be they from the village or anywhere else. Arse pinching she can cope with, if anything gets untoward, you and I will step in. Actually, shall we dance?" I stood up to join the fray but Manohar remained deterred.

Much has been made of the Punjabi dance style. Bhangra is of course a Punjabi folk dance comprising various dance steps which are loosely based on activities common to farming, or at least that's what someone told me many years ago at an afternoon bhangra I attended in Bradford. Patting dogs and changing light bulbs it is not, but reaping crops and ploughing fields it just might be. Anyway, the spirit of bhangra dance is very much 'do what you want' and as it has a heavy floor beat banged out on a *dhol*, joining in really isn't very hard. So I lunged into the fray with an over vigorous arm flailing and high kicking, and presumably *because I is white*, caused quite a stir. As women in India don't really dance bhangra, or are at least pushed to the periphery when the really fast stuff comes on, Julie was still the predominant draw for this touchy-feely dance crowd, but I was coming in a close second. Delighted men hugged and grabbed me, kissed me and tried to guide beer down my neck from bottles tendered across the dance floor while still jumping up and down, shouting 'Happy New Year' and of course "Which country?" as Julie and I reciprocated and replied as best we could.

At one stage as I was doing lots of high kicks, lurching backwards and forwards, shoulders hunched, thrusting pointed fingers repeatedly at the sky, a very stern man stopped me and asked in earnest "How do you know to dance like this?" I can only imagine that in the melee he had picked the wrong man.

After a good hour of this great enjoyment, during which things had got a little too hectic and risqué for Julie as she was joined by two more attractive European girls and the arse pinching got a little out of hand, we fought our way back to the table where Manohar sat perplexed and embroiled in his mobile phone. The shack owner again brought beers and stationed a couple of men around us to ward off the repeated physical invitations of our new friends to come back and dance. I looked around at the jam-packed scene and raised my glass to my host; "looks like you'll be making a few pennies tonight after all."

He fixed me with a serious smile, "It is more than pennies, even pounds possibly I think."

Around 3am we said goodbye to our hosts and vowed to return soon before heading back to Candolim and to our noble hostelry Bom Sporto, via of course, the street omelette man. As we embraced our friends in our drunken sincerity and inflated self worth, I considered what a thoroughly excellent night we had enjoyed. I pondered the possibility of such an occurrence anywhere else and filed it under impossible. No, I shall remain steadfast in my assertion that New Years Eve is amateur night; it was just that on this occasion, we too were amateurs.

Anupam asked me what we had been up to all night and I told him.

"Oh, you know, jumping up and down, shouting, drinking, kissing people, the usual."

"Behaving is too much like amateurs," he said smiling conspiratorially as Julie egged him on, "…surely you is not behaviour as amateurs?"

Chapter Nine

After all the festive fun, the reality of work reared its ugly head. Up to now work had ostensibly been play, but now the real work was about to start, or at least that's how the company saw it. I agreed to spend a sustained period of three weeks at the office during which time I would meet the clients, train the staff and guide the process to finally get under way. I had worked a neat clause into my contract stating that for periods in excess of three consecutive working days, I had the option to bring Julie to Pune with me, and while the company was not prepared to pay for her travel, they would pay her bed and board. I had worked out a cunning plan that if we were to travel together by road or train, footing the bill ourselves, we would be able to travel back together by air and the tab would be picked up by the company. In this way, Julie would not have to travel alone and we would not have to pay for her to fly.

So in the second week in January, we shut and bolted the house, put the car in for a long service and headed off to the travel agent to book the bus.

"Is it definitely comfortable? And both beds are definitely together?"

"It's a luxury coach sir, like the one in the picture." I looked up at the picture of the streamlined Volvo coach with roof mounted wing mirrors making it look like a large solemn insect, a caddis fly maybe. "Beds are together, one this side and one this side."

(Directions in India are often this side or that side, never simply left or right. This extends to beach side, river side, and church side; also front side and back side. The term does not mean *be*side, but rather *on the same side as*. Things are often named after their position in relation to one side of a landmark. According to one writer, the taxi rank behind the airport in Delhi is fantastically named International Backside Taxis. And the term backside, with all its connotations in English is used by everyone with great seriousness, indeed, an airhostess recently inspected my boarding pass as I approached the plane, noted my high seat number and guided me to the rear entrance saying without any trace of a smile, "Up the backside, have a nice flight.")

"What do you mean this side and this side? Do you mean left and right?"

The travel agent shot me a look that said 'yes, obviously' but was somehow at the same time non-committal. She didn't do the head wobbling thing that Indians do, which made me even more suspicious.

"Look, I just want to be sure that this is a very comfortable sleeper bus with good suspension that will deliver us well slept and in good spirit. I also want to be sure that we have got two bunks next to each other, and that we will not have to sleep next to someone we don't know. We have had this experience before, and believe me, on an eighteen hour journey from Bombay on the worst roads imaginable and with no suspension and broken windows, it is not an experience that we want to repeat."

"Sir, trust me, it will be very good sir."

I looked at Julie and she looked positive.

"And you say it takes twelve hours?"

"Twelve hours sir."

"OK, its so cheap, at 300 rupees I think we'll give it a go…"

THAT was my error. How can I not have spotted it? You simply cannot buy a *comfortable* twelve hour bus journey anywhere in the world, certainly not in India, and certainly not for less than four pounds Sterling.

"Hallo, are you ready sir?"

"Hello Peter, yes, here's the bags, Julie's just coming."

We loaded up the taxi with our newest and proudest luggage. Luggage is something that I like to keep in pristine condition. Arrival anywhere with dirty, scruffy, cheap, un-matching luggage is an abomination in my eyes. This is not just about being a Mod; it's rather an issue of confidence. Arrival in unkempt apparel is acceptable and occasionally unavoidable, but one's luggage should display a sense of order, giving balance to one's feelings and appearance. In an unfamiliar location, a confident appearance is often enough to exude familiarity with one's surroundings. Travelling with grubby bags would be like a Galaxy Hitch Hiker not knowing where his towel was, leaving him open to all manner of hostile brigand's attention. A confident case is a useful combatant, helping to fend off those touts and hustlers who prey on nervous new-in-towns. The problem is of course keeping things clean.

We turned into the bus garage to the familiar sound of chanting hawkers and their repetitious bravado peals. Peter pulled right into the station and pulled up in front of a coach emblazoned with stars, suns and other symbolic holiday nonsense, the words Goa Queen in gold on its flanks.

"Can you park here?"

Peter ignored my concern, opened his door and made towards the decidedly not-very-comfortable looking coach.

"Pune, no?"

The affirmative reply came in the form of the ubiquitous head wobble no-yes. I winced, knowing at once my error - *never* trust an Indian travel agent. I stared at Julie who was trying to look silently optimistic, peering enquiringly into Peter's further conversation. She looked at the ground and then at me, tight lipped. She too knew our fate.

Many years ago we had been forced to take an eighteen-hour overnight 'sleeper' bus from Bombay to Goa. At the time the railway was overbooked and the catamaran that we had arrived on was not running for the next three days. Refusing to believe that my status as white foreigner-*sahib* could not buy me a couple of *baksheesh* sleeper beds, I tried throwing some weight around in the train station ticket office.

"Look, surely there must be some cancellations?" I put forward heavily, tapping my fingers on a number of 500 rupee notes.

"Sir, train is full. Reservations are made many weeks ago. There is no room for you. Please go now sir."

No luck there then. I'm clearly no bribe maker. I thought India was supposed to be infinitely corrupt? Whoever said that?

As seems to be the custom in this country, a helpful man had appeared and insisted on taking us to the bus station to secure a passage for us by means of road, as we had been unsuccessful in all other transport methods. I had been very sceptical about his role in all this. He'd guided us some 200 yards to the bus station, advised us as to which bus to take, and what time it departed, and before I'd got a chance to ask him how much he wanted for his unasked-for help, he himself had departed without asking for so much as a thank you. At the station I had asked him why he had intervened and he had replied "It is my duty, follow me". At the time I took his interference as a portent or omen. A sign of a better world in which we all simply helped each other without question and without payment. I later realised that he had been sent by the Devil as a ruse to snare us and give us a premature taste of Hell.

Needless to say the journey was a trauma, an experience, an education and finally an anecdote. It was certainly something to commit to experience. A mistake not to be made twice. Here we were seven years later but seemingly seven years none-the-wiser.

Peter frowned at me. "It's this one sir."

"I know Peter, I know. It isn't supposed to be that one, but I know argument and resistance is futile. As sure as this is India, we're going to have to go all the way to Pune on that fucking shit-bus."

Peter winced, he doesn't swear, but I don't think he was wincing at the swear words. "Just…be careful," he said looking me right in the eyes. He handed me the luggage and got back into his van.

I could see Julie inside our nemesis as I approached the rear of the bus to enquire where luggage would go. I looked grimly at the battered hull; at least it had a lock. I could visualise the inside, I didn't need to look. After some determined investigation I located the on-board *coolie*, a scruffy round boy in a stained T-shirt bearing the acronym 'G.O.A. - Go On Adventure'. I re-phrased it in my mind – 'Get Off At-once'. He opened the lock and lifted the door to reveal a few unkempt cases and some sacks of grain. No livestock at least. I refused his attempts to take the luggage from me and placed it inside myself, as carefully as I could. Happy that it was secure I stepped aside.

"What seat number?"

"Eleven and Twelve."

I shouldn't have said it. I watched in horror as he leant inside and scrawled the numbers in surely permanent yellow wax crayon right across each case. He looked at me and I silently closed my eyes to him. It had already been done, no time to protest. Instead I counted to ten, walked off and boarded the bus.

"An old woman is in our compartment." Julie said. "The conductor seems to think it would be easier if we took these seats. They fold down into beds and at least they're on the ground floor, I don't fancy climbing up there."

I nodded and sat down. Just enough room to put my feet if I poised on tiptoe. I reclined the seat to its horizontal position. "It will have to do."

"Chips, chips, chips, chips, chips." Cried the persistent vendors. "Chips sir? Cashew nuts? Water?"

"No, no thanks."

I counted the seven further efforts of this man to sell me these items before he disappeared, only to be replaced by an almost identical competitor who went through the same ritual with me, but gave up after his weary sixth attempt.

I pulled the stained curtain across the rail and noted with resignation that it did not adequately cover the sleeping space, and that it flapped past the arm of the chair at regular intervals. No privacy for the British. We noted the other passengers fidgeting and walking up and down the aisle, climbing in and out of the seats and beds, fiddling with luggage, repeatedly getting off and back on the bus, and generally getting in the way of each other, while we sat like the reserved Brits that we are, trying desperately to gain some personal space and claim it as our own.

"This'll be alright." I lied optimistically, "I'll be able to sleep, I'm so tired. As long as the roads are ok, I wonder what the suspension's like?"

We started laughing. We could handle this. It wouldn't be too bad. Our neighbour removed his shoes and we stopped laughing.

Without warning, the driver dived into his seat and fired up the old engine. "Great, diesel." I remarked as we started to shake with the motion. Julie clutched her breasts to hold them still. The bus lunged forward and twenty or so passengers who had been standing outside the door attempted to board at once. The driver was waiting for no man. He had travelled some thirty yards before all passengers had boarded, Starsky and Hutch style. The conductor remained resolutely blocking the doorway throughout, passengers uncomplainingly squeezing past him through any available gap. The driver then stopped the bus, presumably to let everyone settle. The bus bounced over a speed bump as it exited the station and we were given an inevitable example of the quality of the suspension.

We were off and I wished we had got off. We discussed our predicament and resolved to acquiesce, determined to laugh it off, enjoy it even.

We settled into our fate like wisened sages or willing ascetics, resigned to our destiny and resolutely putting up with it. Taking it on the chin while keeping a stiff upper lip. We rocked with the bus, anticipating bumps and holes, relaxing into the shaking, heaving, lurching motion, rolling with it, getting into the Zen of it. This was when the hard on happened…

Not a great deal is reported about the illustrious but never mentioned *unwanted hard-on*. We all know what the hard-on is for, and many among us talk about what it is for at every opportunity. What we never mention are the times we experience it without wanting it, or having any use for it. I am not just talking about morning glory, I am referring to those inopportune moments when it is forced upon us unwittingly, times when it is among the last things that we would wish for. Such time might include dull meetings, family gatherings, church attendances and coach journeys. There are countless other examples, doubtless in more excruciating circumstances, but it is not my intention to frighten or upset you, dear reader, merely to point out the frequency and inappropriate locations in which the ineffectual *stiff-o* strikes. It is possible that some of the females among you are unaware of the worthless woody, as you will of course not have come across it yourselves (sorry), and it is unlikely that your macho acquaintances will have drawn it to your attention. Indeed, most men have not broached the subject among their peers; it is known but tacit, experienced but unsaid, felt but un-stated. It is nevertheless very real and very awkward. My theory is that *unwanted wood* is caused by an unwanted situation. It is not, as may be supposed, the jostling of motion that sparks it, rather the degree to which that motion is desired. (Indeed, it is not always associated with movement, what motion is there in the boardroom or at church? What movement in a restaurant or at a football match?) In a situation of uncomfortable, compromised confinement, the irritation is passed from mind to member with the inevitable result. This, to return to the narrative, was my predictable predicament.

"I've got a hard-on."

"No, don't tell me that, this is hardly the place."

"Oh, for God's sake, I don't want it, I didn't summon it, it just happens. This is really too much…"

An hour it lasted. A full Indian hour. The road versus suspension battle abetted it. I fought it gallantly, but lost. After twenty minutes I yielded and just put up with it. There it was, making a tent of my shorts. At least no one could see me; at least I didn't have to get up.

"Toilet stop" yelled the driver. I stayed put, I didn't want to alarm anyone with my triangular shorts, and let's face it, I wasn't going to be doing any urinating. The onset of cold night as we climbed into the Western Ghats finally did for it, and as it receded, the need to pass water replaced it. I had drunk nearly a litre of water in my attempts to diminish it, remembering the words of a friend's father in relation to sexual urges 'Drink a glass of water and run round the garden'. But this wasn't a sexual urge so this was an erroneous remedy.

"I'm dying for a piss now."

"Oh, for Heaven's sake, why didn't you go at the last stop?"

"You know why."

We bumped and surged our way for a further thirty minutes before the bus pulled in to a stop. I rushed to the toilet, following my nose, while Hindi film music blared from incapable speakers. We sat down to eat.

A *biryani* at 12pm. Probably not a good idea, but I was happy to resign to fate. Take the Hindu approach.

"Can I have a lemonade with that please?"

"No lemonade sir, orange will do?"

"*Chalte hai*."

'*Chalte hai*' is Hindi for "so it goes', or 'that'll do'. It's among the few Hindi terms or words I know. The idea of 'That'll do' is so prevalent within the Indian psyche that it is an extremely well used term, known throughout India, even where Hindi isn't spoken. Just take a look at any repair job in the country. Look at the way paint drips are left, or wiring is botched, see how a doorframe is fitted after the plasterwork is done, notice how a new road is finished and then dug up again to fit pipes and cables. In fact I sympathise with the gaffing Duke of Edinburgh when he pointed out that an old fuse box in a factory near Edinburgh 'Looks as if it was put in by an Indian'.

We ate and boarded the bus. The driver set off and plunged us into darkness. Presumably we were now supposed to sleep. We tried to sleep; we really made an effort, believe me we tried. I had to go to work the next morning, sleep was important. Then the cold set in. We had come from Goa, 35 degrees centigrade, and I'd been to Pune twice in the preceding two months. The temperature wasn't much different... wrong.

We were dressed in shorts and T-shirts and had bought no blankets. We had expected air conditioning and the blankets that accompany it. There were no blankets. We also hadn't bargained on the altitude.

"I bought a sarong' Julie ventured."

"Fat lot of good that's going to do, don't be ridiculous."

Julie placed the sarong over her and I watched with envy for at least twenty minutes before caving in.

"Do you think it will stretch over both of us...?"

We fought over the tiny silk sheet for the next four hours. I pulled it off Julie's legs, she pulled it back, pulling it off mine, you get the idea.

We pretended to sleep, fighting over the sarong. We grew irritable in the extreme but it didn't help. We eventually relented and sat up to watch the dawn arrive. India is a breathtakingly beautiful country, but not this day. As we lumbered through the mountains toward our destination, we watched the countryside wake up. People get up before animals here. They get up, leave their houses, walk ten yards and do a morning shit by the roadside. RIGHT BY THE ROADSIDE. We could see the whites of their straining eyes; see the passage of their effluence. We gave up on the view and settled back to some more non-sleep.

The bus finally stopped on the edge of the city some fifteen hours after setting off. "Last stop, last stop, last stop, last stop." Why does everything need to be shouted and repeated so frequently? We waited for the inevitable fight through the doors to subside and wearily rose, cold and uncomfortable. We made no attempt to clear up the mess we had made with our crisp packets and sandwiches; we'd paid for this privilege in every respect. The rickshaw drivers shouted at us through the window, fixing us with their preying eyes. I did not have the humour for this.

"Hallo, Hallo, rickshaw, rickshaw, rickshaw, where you go sir? Osho Ashram? Osho sir?"

"Get out of my way, I don't want to get in your fucking shitty rickshaw."

They parted, eyeing my unkempt appearance with disdain. They don't appreciate swearing here and it had the desired effect. I strode to the hull and watched in horror as the luggage was pitched heedless on to the road.

"I'll get my own out if you don't mind. You shit." (under my breath).

Our beautiful luggage was finished, over, no longer beautiful. The numbers 11 and 12 scored in large yellow characters across each respectively. I lost my currently very hard to find good nature. I approached the scruffy round boy aggressively.

"This is a shit bus and a shit bus company. You work for a shit company and you are shit."

It was the best I could do.

He didn't even look my way.

We were booked into the Taj Blue Diamond Hotel as always and I set about angrily hailing a taxi. The rickshaw drivers asked me what I was doing. I told them. They laughed. I did not.

"No taxi sir, Rickshaw only, no taxi Pune not have rickshaw only please get in where to?"

I unravelled the sentence but defied him completely, resolute that a vehicle suitable for turning up at an esteemed hotel would pass soon. I waited stubbornly while the rickshaw drivers watched, pointed and laughed. After an obstinate five minutes I looked at Julie standing sadly, patiently by the side of the road while I practiced my obduracy, and knowing what her silence meant, I motioned to her to get into a rickshaw. Thirty minutes of round-the-house bone shaking hell and we arrived at the hotel, where the porters took our filthy belongings and watched as we were majestically ripped off by the mendacious but indignant shit-machine driver.

Exhausted, defeated and forlorn, we stood hangdog before the smiling receptionist who professionally overlooked our unkempt attire and asked for our names and passports. Julie and I traded glances; we had not bought our passports. I always forget my passport, particularly when I don't travel by air, but it is essential for checking in to any Indian hotel if you are visibly not Indian. This irritates the hell out of me because I know it is just red tape and nothing is ever done with the forms onto which the details are laboriously transcribed, but rules are rules and we had a problem on our hands.

"I regret to inform you that we have not brought our passports with us, we are resident in India and have left them at home. I know you need them for your forms, will it be possible to just give you the passport numbers that I have written in my notebook and base things on trust? Otherwise we have a real problem."

"Yes sir, that will be fine."

Not looking my gift horse in the mouth, or the eyes, I produced a notebook and flicked through the pages settling on a very empty one. Shielding the empty page from the receptionist I copied two random nine digit numbers onto the forms, put the notebook away, handed one form to Julie and filled the other in myself. They were processed and we were handed our room key. What would the Foreigner's Regional Registration Officer say?

In Gentlemanly spirit I allowed Julie to submerge in the bath first. Her first bath for nearly five months. As she emerged pristine, immaculate, a huge shining whistle, I discreetly ignored the bath of tea she had brewed with her bodily infusion, rinsed it and filled one for myself. My own plunge produced something more akin to weak gravy and I was grateful for my tactful prior silence. The bath transformed me into another man, a bigger, better man, a man who could deal with things, a man you could do business with. I charged a great heap of ironing to the company tab, and dressed in sharp creases, set out as the New Corporate Man to meet my vocational destiny.

I met my new clients at the office waiting for the lift, it had to be them, it just couldn't be anyone else. I only knew their names, Chris and Marcia, but they were decidedly palpable by their size and the colour of their skin. I had spoken to them both on the telephone but had no idea that they were both black, and even less idea that they were both huge. Some forty stone between them easily. Black people are conspicuous by their absence in India, it is hard enough dealing with being white and medium sized, but being black and enormous must bring quite some celebrity and I could see they were not really enjoying their new renown. I introduced myself and asked a couple of bewitched bystanders to remember their manners and wind their necks in.

"Apologies that you chaps are causing such a stir, Indians enjoy a far less developed sense of social grace than us Brits and people such as ourselves are somewhat rare in these parts. Shall we go up?"

Neither laughed or looked remotely diverted by my jollity.

As the lift opened the usual deluge of human traffic spilling forth from the overstuffed box stood stock-still and stared in wonderment at our unexpected gargantuan and uncommonly coloured presence. Beckoning them out with swift annoyance I knew I would have to somehow alert my colleagues to avoid a similar reaction from our client's newly assembled staff. First impressions are important after all, and remembering that as a young boy my younger cousin from Ireland screamed the house down on setting eyes on our Jamaican neighbours for the first time, I felt the need to act. So with a political-correctness-gone-berserk notion I showed our clients into a meeting room and went to alert my colleagues of their blackness and hugeness. As a white man, and a none too tactful one at that, I wasn't sure quite how to convey to my brown colleagues that their new bosses were black and enormous, and that they should display no visible signal of alarm or over interest. They were waiting for me pre-assembled in a training room.

"Right everyone, our clients have arrived and I'm going to bring them over to meet you all soon. There's just one thing. I would ask you to observe some decorum; since they have been in the office they have been stared at rather rudely by pretty much everyone they have met on account of the fact that they are both black and rather on the large side. Oh, and please don't be afraid of them, just be natural, just be yourselves."

"Are you saying we should be frightened of them because they are black?" This was Chandrak, the smirking, smart-arse. He had me on the back foot and he knew it.

"Err, no, no. I was implying that you shouldn't be scared of them as *clients*, not as black people. Oh God, look, that came out wrong, what I mean is…"

Sarasvat lent over to me smiling. "Nice one Mick, great management skills." He laughed. "We have seen black people before you know."

"OK, OK, look, just forget I said anything right, I'm off to get them."

I left the room to mix of laughter and scorn, painfully aware that only white people can truly blush pink.

The first two days of training went very well, structured and scripted as they were. The first real test came when the staff were to take to the phones for the first time. The twenty or so calling staff had been recruited before I had become involved and to be very frank, I had my doubts about their suitability for the job. In my experience, making outbound calls is a job best befitting of people with little or no apprehension about their task and those with a very thick skin. It seemed to me that the staff we were training were far too trepid, possibly even hypersensitive. I hoped that their fears were borne of too much talk and not enough action; they had been kicking around in training for some three months due to the false starts of our clients and some woeful planning from the management. I made a mental note to get involved in the hiring process; we would certainly need some more valiant troupers at the coal face.

The first day calling was nothing short of embarrassing. The staff were making horrifying errors at every turn. Since my last visit they had had no further practice and were hideously under prepared. They were not conversant with the software (a package largely unchanged since its inception back in 1985 in the days of ones and zeros and Doctor Who style computers, not an Indian thing, far from it, more a laziness on the part of the American developers in a very unsophisticated industry), they could not adequately understand the people they were speaking to (to be fair everyone's first exposure to the West of Scotland can be like this) and they were still getting stuck on the "Hello?", "Hallo?" thing.

I was quite candid with the clients about my fears and tried to reassure them that if anyone 'didn't make it' in the job that we would certainly be able to replace them with more suitable candidates. The clients themselves seemed surprised at my frankness and were possibly a little disturbed at my lack of faith in my staff. I reminded them that the company strapline was 'Broadening your Business' and that as a customer they were not just buying outsourced work but would be expected to fully invest a good deal of effort in the staff they were hiring. I asked them again whether they themselves would hire a similar team in the UK. They evaded my bluntness by telling me they had every conviction in the Indian worker as they themselves employed many people of Indian origin in the UK. Political positioning of the correct kind. I bit my lip and let it go, but I felt my first inkling that the regularly patronising 'niceness' of these clients was so immersed in fear of discrimination that they were terrified to say what they really felt.

It was clear to me that these were not the people at the client company who had actually taken the decision to outsource their work to India, and that in their very operational role, were unaccustomed to the financial aspects of the big bad world of business. It occurred to me that far from being the people who wanted the new relationship with India to work, they were instead the very people whose jobs would be replaced were the affiliation to prosper. I decided that it would be in my interests to know what they really felt about the whole affair and decided to do what any self-respecting businessman does when the need to ask direct questions arises, and get them drunk.

We had been out to dinner a few times during the first week of their two-week stay. I was quickly finding out that Indian business hospitality was very much akin to the generosity of Indians the world over, that is to say munificent. They were being put up at Le Meridien, the most opulent hotel in town (with the worst Indian restaurant I have ever encountered) with all expenses paid, they were being ferried from pillar to post in air conditioned vehicles and were being showered with gifts at every opportunity. And I too gushed at them as to how wonderful and different India was in an effort to impress on them the possibilities inherent in placing their business with us. I realised early on that my own genuine enthusiasm for the country would do as well to sell the business as anything else I could do. A true advocate with relevant experience from overseas among the indigenous talent to prove how fantastic the place is, is a great marketing tool and I intended to use it. So we remarked together on the prodigious guivor of the side-saddle riding, sari-clad motorcycle pillion passenger women and the impossibility of negotiating anywhere other than an Indian road on a bullock cart. We jointly enjoyed the occasion when a painted elephant lumbered its way along the main road before the office and we expressed dual amazement at the flock of babblers pecking violence and bewilderment at the smoked glass windows of the office top floor.

But the time had come to interrupt the pleasantries with some guileless investigation as to how they really felt about things. Five of us took them out to a nightclub. They were both over thirty and had displayed very traditional viewpoints on a variety of topics thus far, much befitting of their years, but they were both from Jamaican families and I figured that there might be a more relaxed persona behind both facades, maybe if we could get them in a more relaxed environment we might see a little more of another side.

I was very wrong. We paid their cover charge, dispensed with the obligatory Indian bar snacks and set about the booze. Or more precisely, I set about the booze while they watched in horror. While drinking I told them of my early exposure to Jamaican culture via my indomitable neighbours and school friends, how I had first smoked cannabis in the lounge of a staunchly black pub with old men playing dominoes, and how I had been the very epitome of 'black culture white youth' but had never had a black girlfriend despite my efforts. I was incorrigible. Slowly and surely I trotted out every single cliché about black people that came to mind, even ordering a round of white rum as a toast, and by the time I realised how embarrassed we all were by my astonishingly dumb performance I had dug myself too deep to recover the situation myself. Shanti came to my rescue in a way she would repeat over time using three simple and direct words, 'Shall we dance?'

And dance they did. A full hour on the dance floor; most probably just to escape me.

Two days later I took them out to lunch on my own. I didn't drink and neither did they, and I just asked them outright how they felt about the whole Indian outsourcing thing, and they were much more candid than I had expected. It transpired that they had very different views to each other and I deciphered that they were probably not very good friends. Marcia was fairly open to the idea but had reservations about the communication aspects, her fears were that they would struggle to control what went on in so distant a location, but she was impressed by the company infrastructure and very sure that with guidance we would be able to meet their expectations in full. She noted that the staff had improved immeasurably over the two weeks and were more or less at a level she was happy with, and that the management team seemed very capable. She concluded by saying that she had enjoyed her stay enormously and was very much looking forward to coming back. Chris was far more sceptical. He whinged about the lack of experience in the staff and the fact that everything started late and that he had never been picked up from the hotel on time. He moaned about the country in general, an overall lack of professionalism and inability to communicate on his terms. When I pushed him he eventually blurted that he felt the whole idea of taking jobs from the UK was doomed to failure and it would only be a matter of time before the whole nonsense blew over. I nodded seriously throughout, given my performance a few nights before I felt it to be the best policy by far, and we finished up our lunch and drove back to the office quietly observing the Indian world go by.

The next day they left amid huge smiles and great promises. I was sure that we had done enough to win the business in the long term, despite the varied doubts of the people who had just left and my own outrageous attempts to perturb them through racial stereotyping. I just hoped that the next clients would be a little more senior in their business life and thus experienced in the more pertinent financial reasons for doing business with us. I also prayed that they would not be Irish, Scottish, Welsh, Anglo Jewish, Afro English, or from any other walk of life to which I feel I have some claim and thus the right to comment unstoppably upon. Bring on the Americans I thought, there is nothing I can do to offend them.

While I had been immersed in my occupational *faux pas*, Julie had been industriously looking round the city of Pune, and despite a seeming lack of anything to do with the art, she quite liked what she saw. She reported back from the Raja Dinkar Kelkar museum's most wonderful collection of predominantly small things in a most amused manner, saying that above all in the eclectic compilation of just about anything the great man cared to pick up during his life, the sundry zoological vegetable choppers were easily the most memorable. She scoured the city for shops and found a far greater range of commodities than in Goa – supermarkets, department stores, chain restaurants, all the things we came to escape but secretly missed. Together we visited the Gandhi National Memorial in Yerwada where the Greatest Man Who Ever Walked The Earth was interned with his wife Kasturba for just under a year while the Britishers panicked over the Quit India Campaign and incarcerated the entire Congress government. Of the copious monuments to Gandhi I have visited it is one of the most moving, seeing a pair of his spectacles and a pair of his simple wooden sandals conferred to me a tender connection to the man whom I cannot even think about without emotionally welling up. What a man.

On Julie's recommendation we avoided the Osho Ashram, Pune's world famous 'Meditation Resort' founded by Bhagwan Shree Rajneesh the infamous sex guru who had been deported back to India from the US in the 1980's following a charge of immigration fraud which was most likely pinned on him by the Reagan administration who were getting worried about escalating rumours of the mad sex sessions at the ashram in Oregon. Today the Pune ashram is awash with maroon robed self seekers from around the globe who pay top dollar to enroll at the 'Multiversity' to study everything from 'Craniosacral Balancing' to 'Being a Man', or simply 'Reminding Yourself of The Forgotten Language of Talking to Your Mind and Body', which, let's face it, we all need reminding of.

****

No sooner did one new client leave than another arrived and I was embroiled again into work. Given the amount of time things were taking, and the fact that after two weeks there was little else to do for a single white woman who doesn't particularly enjoy being gawped at by sexually frustrated men in the street, Julie set off back to Goa alone where I would join her in a week or so.

My next client was Steve from a huge American conglomerate who wanted tpo place an inaugural five staff to see how things went, and all being well, paying for a whole host more. A senior company director with a wealth of experience in offshoring, a novice he was not. Not for him my simple enthusiasm for all things Indian. I figured that my cute pandering to how wonderful the seeming madness of India was would not be sufficient to win his business; I would have to rely on the efficiency of our operations, and the ability of my team to get the prescribed work done, about which I was still more than unsure.

I needn't have been concerned. We had hand picked the five people from the larger team to give him the best ones and Steve was mightily impressed when he met them. He trained them for a couple of days and they started work. It went extremely well, far better in fact than projected. After two further days he had seen enough and signed up for us to start training a further ten people. It was really that easy. As I had suspected, the difficulties we had experienced with the rest of the staff were indeed down to unsuitable hiring rather than any cultural gaps. It was simply a matter of finding people who could do the job, just like anywhere else in the world. Steve turned out to be very valuable to me in my learning about the whole call centre/outsourcing field, as head of outsourcing for his company he had offshore suppliers throughout Asia and South America and was very sure India would eventually corner the English speaking market due to its superior education system and instilled business acumen. He also boosted my own confidence by commenting on how good the staff we had assembled for him were (in time this team would go on to become more productive than any other group of people I have managed).

Just before he left for China with a promise to return soon, Steve and I were having a beer at his hotel when he asked a favour of me.

"Would it be possible to arrange a driver to take me to the Osho Ashram tomorrow? Don't look at me like that; there is a good reason. My boss is from Oregon and he has a photo of the Osho place there, which he is very proud of, particularly as you are not allowed to photograph the place. I thought it would be a good laugh to give him a sister photo of the place here in Pune. All I want is for the driver to drive past while I snap a couple of shots, I went past yesterday and took my camera out and the security guards came running towards me waving their hands, so I'll need to be quick. What are the chances?"

"How can I refuse? One condition though - I'm coming with you."

I arranged for the driver to pick up first myself, and then Steve from his hotel from whence we'd do the deed before whisking Steve off to the airport quick. I booked Satish, a driver who demonstrates a rare driving skill, namely the ability to abstain from using the horn. We picked Steve up and headed off to Koregaon Park and to the ashram. As we approached the Meditation Resort gates I could see the security guards scouring the streets, a very rare vigilance for normally underpaid and under worked hands. Satish knew he had to drive slowly and steadily enough for Steve to get a good snap, but quick enough not to draw unwarranted attention. As planned we drove past once as reconnaissance to determine the whereabouts of the guards and upon coming round the block and into the main road for a second time, a ridiculous sense of anticipation and fear spread between us, manifesting resolutely in giggling laughter. Steve readied his camera in the back seat as I knelt on the front seat leaning into the back to wind the window down and quickly back up again. Satish's giggling reached squeaking pitch as we made the sentry post and I began winding furiously. Steve's booming laugh leapt from the window alerting the guards and as they looked directly at us he fired five or six snapshots in rapid succession. The shutter noises, Steve's rumbling bellow, Satish's piercing screech, my own laughing and the over-revved engine produced such a bewildering dissonance as to petrify our sentinels to fossils of their former selves. As we drove away they looked to one another and shook their heads in 'Surely that didn't just happen?' solidarity and dismissed the whole episode from their tired minds.

By the time we had calmed down and were well on the way to the airport, we examined the photos on Steve's digital camera. The pictures showed three utterly perplexed security guards waving hands in front of a huge sign saying Osho International Meditation Resort. Steve was delighted and I knew that sharing such a preposterous episode would be well-invested money in the bank to secure against any future mistakes by myself on the work front.

With Julie gone and the clients gone I spent three more days in the office before returning to Goa. During these days I reflected on how much work there was still to do, despite Steve's impressed eulogising. I also reflected on how much I had enjoyed the experience of working again. I began to consider how useful, even *essential* my contribution would be to the success of the whole thing, and I had already begun to feel responsible for my new workmates' ill-considered careers. Maybe there would be a way I could spend more time at the office, more time helping out the needy in this way? Maybe this could be my big contribution to India, my way of helping in the community? Come to think of it, Goa was going to become very hot soon, and then the rain would start and the whole place would shut down, our infrastructure of bars and restaurants would vanish and we would be holed up in the house with little to do, increasing power failures and no English football to watch. Maybe we could move to Pune for a few months? Julie liked it after all, even if there was no art community as such, she could surely find something to do there? Maybe I could get the company to pay for it? Give me a full time job for a few months, just while I helped get things properly off the ground?

My mind raced with the work drug and a dirty new seed was sown in dangerously fertile ground. Like a junky I became blinded to the trap, trotting out all the clichés 'it would only be for a few months' 'what harm could it do?' and of course 'the money wouldn't go amiss'. Like the 1970's Superman Doesn't Smoke campaign, I was at the mercy of Nick O'Teen 'Just one puff and you'll be in my grasp'. Fatal.

I flew back to Goa hatching my plan. I was convinced by now and was sure I could convince the company. Julie was the only obstacle. It might take a little more than the promise of a bath this time.

Chapter Ten

In Goa the weather was becoming far more comfortable. December to February is most pleasant, with less humidity and long sunny days. We had invited numerous friends to arrive in succession for a holiday, recommending January and February for the climate. Three couples came over a in one month period bringing with them newspapers, sausages, stilton, Twiglets, Sarson's malt vinegar and the holiest of holies, four bottles of Fullers ESB.

It is often said that no sooner do you have a place in the sun than you have a place in the sun full to bursting with unwanted overbearing friends and relatives, guests who neither pay their way nor know when to leave. It was quite the reverse for us; apart from my father we had seen nothing of the countless friends who had assured us they would visit, not hide nor hair of those who had been vying for space in our house by the sea before it was even properly conceived, and rather than be delighted with this solace and solitude that we had sought so fervently, we were actually rather desperate for a bit of familiar companionship, someone with whom to share the fun and madness, someone to show the place to, someone to confide in, someone to properly laugh with.

First Max and Mandy came, extraordinarily well travelled for a married couple, having spent the best part of the last ten years flitting between job hate and wanderlust. But India's charms had largely eluded them. They had spent a few days in Delhi some years back on the way to Nepal and had found the aggravation insufferable. Max had recounted to me an incident where a street shoe mender had literally dragged Mandy to the floor in an effort to re-glue a very slightly flapping sole, screaming "but madam I can fix it, I can fix it" while enduring her well aimed kicks and stoically hanging on. The sheer quantity of personal interruptions, insistent beseeching and occasionally plain weird street hassle had sent them scuttling north, forgoing their proposed two month stay entirely, never to return to Indian shores. Until now that is. I had given them my usual money back guarantee, stating that if a trip to Goa did not render null their opinions, I would refund the full cost of their holiday. Being from the downsizing fraternity and honorary members of the New Skint, this was an un-missable opportunity for them, perhaps even more attractive than the 'You've won a free holiday, all you have to do is come and listen to a presentation on timeshare offers' proposition that is so common in the areas of southern Europe which serve bacon and eggs for breakfast. More attractive maybe, but no less of a con. In my enthusiasm for India I was certain they would enjoy its fruits, but I was also acutely aware that anyone's definition of a good time is a very easily contestable point. If it didn't work out, I would quickly resort to semantics.

Rather than stay with us, we had actually asked our friends to take package holidays with hotels in our vicinity on the basis that the house was small and that we didn't want to get lumbered with them if they got sick. Most people get ill to some degree when they come to India. The British refer to this so hilariously as Delhi Belly of course. For the most part it is the water that lays people low with mild nausea and diarrhoea, usually they have brushed their teeth with unboiled water or eaten a salad washed in it. I have had various bouts with the same but am largely unconcerned by it, it lasts around four days during which I crap a lot and moan a whole lot more and when it finally stops I am usually pleased to have shed a couple of pounds which I can thoroughly enjoy regaining again. Many people however, seem to be unusually perturbed by it, insisting on the presence of a doctor, a multiplicity of medicine and a twenty four hour 'please keep checking that I haven't died' vigil. It is this duty of care that I had no wish to extend to my friends, the hotels were used to it; let them demonstrate their expertise.

Max and Mandy were perfectly happy with their hotel and after three days were utterly delighted with their holiday experience. The cheap excellent food, paradise beaches, exhilarating rides in our comedy automobile and the sense of palatable foreignness that Goa so effortlessly provides had worked their charms and they were already expressing an interest in exploring India further, as I knew they would. When Lucille and Jim arrived however, things panned out a little differently. A pair of untravelled, unprepared and unlikely adventurers, they arrived at their accommodation-on-arrival hotel exhausted and perturbed by the wretched sardine tin charter flight and the carefree driving style of the transfer bus driver, to find a dingy miniature room awaiting them with a half door that would not admit their luggage, let alone Jim's six-foot-six-inch frame. It contained a stridently buzzing fridge with the last guest's half eaten Toblerone still partially intact, an eye catching view of an electricity sub station and a family of mosquitos casually lounging on the scruffy little bed saying 'come, lay down, let us drink your blood'. They complained and were told they could 'shift' to another room tomorrow. The new room was not forthcoming and by the next evening the mosquitos had called in their friends - 'hey, you lot, there's a couple of white ones in here, fresh as daisies, come and join us, there's plenty to go round'. I joined the complainants and located the holiday rep. She was easy to spot in her yellow and blue shirt and yellow and blue eyes. Like all holiday reps she was part alien part robot and spewed forth the usual empty promises before quickly clambering her fatness onto a moped and scooting off back to her dream boyfriend, a none too fussy local man with an unconcerned redolence, a wife, and an eye on a British passport. We gave up and booked them in to the consistent Casablanca Hotel for a few days of parasite free recovery.

After a week, everyone had settled down and we planned a trip together. We had agreed that Goa was not 'Indian' enough to suffice as the 'Indian experience' and that we should travel out of state to broader horizons. By popular consensus we decided to go back again to Hampi. Julie and I both felt that we had not really seen enough of the place with my father in tow and fancied staying in the village itself for a few laid back days cycling round the ruins and messing around in coracles on the river. Rather than repeat the bone shattering road journey, and given that there were now six of us, at least one of which was the size of an American, I suggested that train travel would be the best option and tracked off to Panjim to get some tickets. Train travel is excellent in India; at least I think it is. Yes, it is painfully slow, extraordinarily crowded and while at stations at least, very very smelly. Yes, the toilets are a hole in the floor, the seats are hard and the perpetual stream of hollering vendors selling, 'Chai-ah, chai-ah, chai-ah' or 'Coffee, coffee, coffee, coffee, coffee' throughout the day make sleep more than difficult. And I know that the railway system offers neither French speed nor Swedish efficiency, but I maintain it is still excellent. The tracks are flat and smooth in stark contrast to all other forms of Indian travel and there is just something really enjoyable about the whole experience, a sense of friendliness and camaraderie among passengers that you simply don't get anywhere else. There are various standards of carriage, from first class air conditioned sleeper, where you can expect sheets and blankets (you'll need them), adequate bed/seats and a smiling host to give you breakfast, lunch and dinner; to third class chair cars where you will have to fight for an inch of wooden bench and be stared at continuously by emaciated toothless oiks who will not speak a word of English but will nevertheless offer you breakfast, lunch and dinner that they have brought with them in well organised tiffin tins which, without exception will be absolutely delicious. I felt my friends should experience something between these two extremes and bought second-class non air-conditioned sleeper tickets.

Buying train tickets in India, in fact buying pretty much anything in India, is a bloody hassle. Depending on the size of the station, there could be any number of bureaucratic steps to complete before you have your ticket, and even then it is not always clear that your ticket is either the one that you wanted or the one that you paid for. In Panjim the station is fairly small and the ticketing procedure is at least completed within one office. It does however involve buying a ten-rupee ticket containing nothing other than a number that may or may not appear above the one open ticket window or the four closed ones. If you are lucky enough to see your number appear, you will have to fight for service amid ten or so others who have a completely different number to you or very likely have no numbered ticket at all but nonetheless crowd, jostle and shove you in the back continuously while you are trying to explain where, when and how you want to travel, to the indifferent member of staff who is half cocking an ear to your request and half having a conversation with their docile colleague at the next window, who is also not listening, being far too busy doing nothing at all. If you are able to communicate amidst the bedlam, be prepared to compromise on the cost, class, date and even destination you are trying to travel to, as trains are usually booked some months in advance. On this occasion, after an hour wait, I was lucky enough to secure tickets with the exact date, class and destination requested for all six of us, at least for the outbound journey, the return journey was 'waitlisted' even though I was handed (and paid for) the tickets. I was told I would need to confirm our intention to travel one hour before the train arrived at the station to see if the seats had indeed been allocated to us. This might seem dangerously *gung ho* but experience told me that, unlikely though it might seem, the seats would be available and our names would be posted on the side of the train carriage we should board. I knew too, that if they were not, buying someone else's seat from them was always a possibility, failing that, the one inch of hard bench beckoned, but whatever happened we would get home.

At dinner the evening before departure, Jim began to complain of stomach cramps and we all exchanged worried glances. A man with his ingestion talents rarely spurns food and as he declined to order we knew something was very wrong. He sat in silence while we imbibed, ingested and enjoyed. As the desert arrived he could sit no longer and, scraping his chair awkwardly back, he exited for the rest rooms with a lumbering gait which we remarked was most befitting of his nick-namesake Grape Ape. After far too long in the toilet, the Incredible Hulk returned. His clothes were intact but he was green, enormous and sweating.

"I must go now" he asserted. We didn't argue.

In view of the 6am start I had arranged for all to stay at our house where Peter could pick us up easily. Within an hour of returning home it became clear that Jim would not be coming to Hampi. Spread-eagled across the spare bed, he had taken on an unsightly green and white pallour, his eyes had become small and distant and his mouth was doing all the breathing without the help of his nose. His manner became childlike as he turned from the Hulk into Shrek. Then the puking started, empty bilious puke, the most exhausting variety. The rest of us packed our bags and settled to sleep while Jim lurched between bed and bog. We tried to sleep but his exertions ensured a bothersome night and it was with little sympathy that we took our leave of this helpless hulk, leaving him in his odorous fugue, telling him to drink plenty of water and get some Imodium. As we would be away four days, I tried desperately to talk him through the electric water pump system that he would need to operate if he wanted more than one shower, which I was certain he would. In his delirium I could not fully convey the necessary instructions, but I was sure he had at least understood that it would need turning on. And thus we rather cruelly left our laid up leviathan to his sickness and departed for some Indian Railways fun.

Whilst I am an advocate of Indian Railways, it is fair to say that my compadres were not overly enthusiastic. I don't think the fact that our bunks were occupied by fare evading vagrants helped. Once the guard had ejected them and we had sprayed some mosquito repellant around, I pointed out that the smell wasn't too bad and settled in for a ten hour stretch. After an hour or so the roles had reversed and in my efforts to get some sleep, I had become irritated in the extreme. I tried to communicate to my friends that their over politeness with the stream of insistent hawking vendors was only confusing matters and that unless they intended to buy something, any conversation was pointless, conversations such as...

"*Bhel puri*, biscuit, lollipop, *bhel puri*, biscuit wallah. Biscuit madam?" (Shouted into their observing faces).

"No thank you, I don't want anything thanks."

"What you want madam (hollered into the ear) *Bhel puri*? Biscuit? Lollipop? *Bhel puri*? Biscuit? Lollipop?"

"No, nothing thank you."

"*Bhel puri*? Biscuit? Lollipop? *Bhel puri*? Biscuit? Lollipop?" (Thrusting a large bag of oversized rice crispies into Mandy's face).

Or perhaps:

"Chai-ah, Chai-ah, Chai-ah, Chai-ah, Chai sir? (Pouring a cup at the greeting nod of a head) Chai sir?"

"No, I didn't order anything, no, no thank you."

"Chai sir (shoving the little plastic cup forcefully at Max), chai you want? Five rupees, five rupees, (angrily) give me five rupees."

...could be avoided by not looking up and smiling each time a vendor walked past. I also asserted that the screaming child in the next compartment was certainly not cute and the game of 'where's daddy? I don't think he's ever coming back' which its mother found so amusing to torment the child with while her husband was in the toilet was not at all endearing, even now that he had returned and loudly reunited himself with his offspring. I realised that getting any sleep was just not possible during daylight hours, so after a further hour of futile huffing and puffing I relented and joined the others in marvelling at the world outside the window. As we stopped at Dharwad we jumped off and joined the queue for the best masala omelette any of us had ever tasted, the combination of roughly chopped and raspingly hot chillies and the newspaper print rubbing off onto the fluffy, greasy omelette from its receptacle conspired to create something that we were unanimous was in fact Manna. The addition of 'lunch' comprising plastic bags distended with lipsmacking vegetable *biryani* put us into superlative mood despite our half-slept fatigue.

As the afternoon wore on I suggested we spend some time sitting by the open doors surveying the land and watching it pass by. An upshot of the absence of safety measures in India is that very often common sense is left to prevail. In this way we are not harassed by signs telling us to keep the doors closed, not stick our head out of the window, not to stick our head in the oven and not to jump off a cliff. Instead, it is left to the individual to decide whether to risk sitting by the open door and smoking a cigarette or not. You fall out, that's your problem, just don't blame us, and don't expect the train to stop either. I love the freedom this extends to us, and I love just sitting there in the shade watching domestic scenes unfold before me; trading waves of the hand with smiling rural folk and listening to the quiet roar of the train's horn as it rounds bends, warning the variety of beasts of its relentless encroachment; this is travel as it should be, relaxed, slow and contemplative. I suggested my guests spend some time acquainting themselves with the beauty of this astonishing land, sat them at an open door and left. An hour later Max and Mandy returned with familiar tales of village life and we mused as to whether gossiping around the village water pump was akin to tittle-tattle around the office water cooler. I was reminded of my plot to get myself *back on the work* in a full time job and took the opportunity to air the idea in public. I told my friends how I felt there was an opportunity for Julie and I to stay on in this first-rate land for a further six months and to extend our overall stay to eighteen months. I enthused about how taking the job might fuel my dormant writing and how if nothing else it would give us somewhere cool and dry to sit out the monsoon and get paid for it at the same time. Julie glowered a little but otherwise remained cool. I felt I had done a pretty good PR job for a first attempt as everyone agreed that it was an interesting option at least. I asked Julie outright whether she would consider staying on a further six months, she was non-committal which I considered a victory. It would now just be a matter of wearing her down, a despicable strategy but a successful one nonetheless, a line of attack I had used as a child - just keep asking until they say yes.

After another hour or so Lucille returned from staring out of the train door. She carried a distant smile of satisfaction, elated, enlightened even. The magic difference of things had got through to her and it showed. She sat down with a contented sigh.

"I have never felt anything like this in my life, just sitting there I really felt like I was in another world. This place is unbelievable. It is amazing. I love it."

We alighted at Hospet as planned and surveyed the scene. The anticipated mob of revving auto rickshaws was supplemented by a handful of hopeful looking cycle rickshaw wallahs, all vying for the most lucrative looking foreigners to take to Hampi. There is one train a day in this direction and almost everyone getting off at Hospet would be going to Hampi, some 12 kilometres away, so this was a one shot deal for these drivers and the desperation was very much in evidence. I quickly quashed the interest in the quaint cycle rickshaws and guided my cohorts toward the autos. I normally can't be bothered to bargain for rickshaws, choosing instead to just get in, state my destination and pay at the other end, (you never see James Bond arguing over a price) but here it was clear that a rip off was imminent. After a bit of haggling we settled on an inflated price and took two rickshaws for the five of us, Julie, Lucille and I sharing one, Max and Mandy the other. The driver we reached a deal with walked away from the decent looking vehicle he had been leaning on and returned with a broken husk of rusted metal, pushing it hopefully in our direction.

'No, No, No" I objected, "We're not getting in that, no chance."

All around us people piled into black and yellow rickshaws and raced away, buzzing like clapped out bees in a race to the hive. Max and Mandy's chariot too sped away, leaving us with little choice than to jump into the desperate wreck before us and hope for the best. It was clearly a race to the Hampi hotels. As any user of the Lonely Planet travel guides will know, making friends at airports or train stations is a bad idea, instead, one should be first away and running for the taxi rank to get in front of everyone else who will be heading to the best cheap hotel mentioned in that unavoidable guide book. The race was on.

Being three fully grown adults in the back of the oldest and worst cart on the track didn't help our plight any. We set off with Lucille on Julie's knee but soon realised the weight was not counterbalancing the three wheels and we were in real danger of only going in circles. We redistributed as best we could with the two women sitting on the seat and me centrally on their laps, put the bags either side and tried again with more success. The state of the vehicle and weight handicap did nothing to perturb our indomitable driver and we soon realised his gritty resolve would see us in better contention than expected. He sailed through the gears at insane speed and baring his tannin-tinged teeth to the wind, jammed his hand underneath the throttle and opened the rattletrap up.

We had been about seventh or eight off the grid in the most pathetic crate in the bunch, but to our amazement we were gaining on those in front with real momentum. First we dispensed with a couple of women bedecked in the classic foreign traveller getup of tie dyed everything who had the hangdog look of unconcerned resign as we passed with our arms aloft. Next we successfully passed two young Englishmen who we jeered in a very un-English way much to their discomfiture. At the first set of lights we sailed through heroically taking third place past two drivers who had stopped and turned off their engines, their longhaired hippy passengers shaking their matted mop heads in displeasure. Suddenly trouble reared, Lucille noticed that the left rear wheel was wobbling badly and suggested we stop and check it was ok. I was having none of it and neither was the driver. I think he had his mind on getting back before the others so he could claim a better place in the queue at the station, but I was giving attention to nothing but victory.

"Mick, I am worried about this, please can we stop?"

"No way, we need to get there before the others to get a hotel room, it will be fine, don't worry."

Lucille looked at the violent wobbling of the wheel and around at the terrible state of the madcap *tuk tuk* and put her head in her hands.

"How far do we have to go in this thing?"

"About ten kilometres further I would guess, maybe a little more."

"Right, I'm getting out." Lucille motioned to the driver to stop who ignored her impossible request and headed determinedly on. She stubbornly swung her leg to the left and took hold of her bag, motioning as if to jump. The balance was devastatingly upset and the vehicle swung dangerously left. The driver shouted something above the whining din of the underpowered engine and wrenched the speeding jalopy back on track. Lucille sat quietly again, defeated, scared, terrified perhaps. We sped onward.

I could see two auto rickshaws ahead of us and in one I recognised the protruding strap of a bag I knew to belong to Max and Mandy. We had a long straight road ahead but we were gaining on them slowly and surely. The driver started making little lurching movements in his seat, willing the vehicle on, and after five minutes of burning rubber and full throttle engine abuse we were within metres of them. And they knew it; they were peering out of the sides nervously giving us the V's as we drew neck and neck. We drove with only inches between us for a full minute, neither driver distracted from his task despite the whooping disruption of our bawled expletives. Finally we took them, the smiles dropping off their faces as their driver made the mistake of swerving to avoid a wayward pig and we pulled ahead and stayed there. Despite the insane efforts of our determined driver, we couldn't catch the clear winner who must have been running a deluxe model, but we were proud in second place and as we reached our destination I gave the driver a handsome tip much to Lucille's unamused chagrin.

We waited for the others and booked into a perfectly horrible hotel in the medieval village centre. The rooftop restaurant looked out over every other identical hotel rooftop restaurant that comprises this strange little place. We had unwittingly timed our arrival with the Hindu festival of Sangkranti, meaning that the place was awash with pilgrims doing what they do whenever they come across a river, washing. In the morning we took bikes to the river crossing and witnessed a good two thousand faithful either washing or doing their washing. The prospect was an Indian cliché the like of which I had never actually seen, thousands of semi naked people swimming, bathing or standing around on rocks drying, their colourful robes and saris strewn about them also drying in the sun.

We spent two days in Hampi while the festival buzzed around us. Scooting about on bicycles visiting the ancient sites, secretly banging out tunes on the hollow pillars of the Vittala Temple knowing damn well we shouldn't, climbing to the Hanuman Temple and exhausted, wishing we hadn't, clowning around with the monkeys and generally having good clean fun. Oh, and having our pictures taken of course.

Many places you travel throughout India will throw up people who inexplicably want to take your photograph. There must be hundreds of people from all walks of life who have photos of Julie and me in their albums. I used to think this was a male opportunistic thing, showing pictures of the female *farangs* they could claim to have had some relations with, but over time I have surmised that this is not the case at all, as we have been snapped by as many women as men. In most parts of the world, it is the travellers taking photos of the natives. In India, things are inevitably reversed. We genuinely must have been asked for a photo over one hundred times in the two days. One can only assume that people in Karnataka don't get out much, because why they should see us as celebrities I really cannot say. To say we were getting stared at by the Hindu Pilgrims is an enormous understatement; we were drawing a constant and unabashed crowd. There were many sideshows on offer due to the festival - people putting swords through their faces, people with long nails, long hair, that sort of thing. But they had nothing on us. At one point I went to have a look at a pinhead woman who was sitting on a rug making an absolute fortune as people flocked to give small donations in return for considering her very small head. As I closed in, daring myself to take a photograph, I must have been clicked ten times before I knew where I was. In fact I sensed some unrest and was sure the pinhead mafia were sizing me up for stealing the limelight, so we hotfooted to a local 'bar' which we determined would secretly sell us beer despite it being both a 'dry day' and party to the restriction on openly drinking in a sacred place. We took a seat and the crowd followed to observe. After some real Laurel and Hardy comedy where the waiters argued as to whether we could or could not have the two bottles of beer that were produced, finally deciding on a teapot and tea cups to serve it in to try to fool the local constabulary should they put in an appearance, we took our first drink observed by no fewer than forty people who stood on tiptoes peeking over the fence, looking like dismembered heads on sticks, staring at us with fixed gawp over the spiked fence posts.

The next day the festival was over and Hampi was empty. No elephants, no Saddhus, no pinheads, just impossible, beautiful Hampi. On the train on the way back we all slept. Our beds were vacant and we used them well. The three nights in a tiny room with uninvited chickens and pigs had taken their toll, and the resident rooster outside each hotel had made sure of very early starts. After ten hours of comparative railway peace, interrupted only to sample the amazing omelettes again (it wasn't as good of course, one should never go back) we arrived at Margao station and after a terrifying ride stuffed into a tiny Maruti taxi van we reached home. We pulled up to the house to find Jim sitting out on the verandah looking skinny, pale and somewhat sheepish. We asked him how he was.

"I have of course been better. The worst thing was the fucking neighbours coming round all the time. Apparently I left the water running or something, you'd think the world had ended, they claim to have been banging on the door for hours because water was dripping out of the overflow, you must have left the pump on. I was fucking hallucinating in there, shitting and puking and shivering, I had no idea about the water. Your landlord shit himself when I finally answered the door, I don't think he is used to people of my height. He was all angry so I told him to fuck off and he soon calmed down, or at least he left in a hurry. Honestly, what a fucking nightmare."

I went round to Arnold's to try to make peace. It seemed he had come to terms with the incident that was by now two days old. He said only "I thought someone was dead in there, he did not need to be rude, he is so big, when he opened the door I thought only of Goliath". I apologised and explained as best I could and finally went to meet the others in Bom Sporto where Jim was braving the first thing to pass his lips other than water in four days - lager. Anupam commented on how much weight Jim had lost.

"You are looking too much of thin no? So much of weight is losing here, what happen? You are being sick in India no?"

"If you call projectile vomiting, shotgun shitting, terrifying paranoia brought on by terrifying hallucinations, lonely fear, loathing and a near death experience 'sick', then yes, I have been sick."

"Is no problem, all white people coming India getting sick. White people having too bad no good stomach. Too weak. Is only sickness. You will feeling better tomorrow. Have more drink, English people needing very much more to drink."

Later that evening I brought up the subject of working full time again. Julie still looked doubtful but seemed encouraged by the reaction of the others. Their own enthusiasm for the country (not Jim of course) even had them pronouncing notions of jealousy, willing me on to take the job and stay longer. I wasn't in need of the encouragement, but I could see it was working on Julie.

Two days later I asked Julie outright, sure that I had done enough tentative idea placing. She just said yes. No arguments. As long we stayed in Pune for no more than five or six months and as long as I could earn enough to cover our costs and enough to help with a further six months in Goa afterwards, she thought it was a great idea. It would enable her to spend more time on her drawing and painting without interruption and it would be somewhere to shelter during the rain. She did however make me promise that after the six months was up I would jack in the job and we'd come back to Goa to do what we had planned to do - write, paint and generally fart about. I agreed immediately. What did I care? Who knows what excuse I could come up with after six months? The important thing was to secure the position, I went straight to the phone.

"Avinash? Hello, this is Mick, how are you doing? Good. Listen, I have a proposition for you. I have been thinking about how beneficial it would be for the company if I spent a more prolonged period of time working on a full time basis. I would like to offer to work for a six-month period in Pune to help get the department off the ground. This is such a crucial juncture that I feel it needs greater input from me and it is going to prove very difficult to manage the thing from here in Goa. What do you think? Are you interested? You are? Excellent, why don't you have a think about it and come back to me with some kind of offer, the money is not really that important, just make sure it is enough to cover all my costs and live like a Maharajah and I'll be happy. I'll speak to you tomorrow then, OK? Goodbye."

In one short phone call I had blown all my efforts to escape and fallen hopelessly, haplessly, helplessly into the abyss. I was right back on the medication, drugged up to my work thirsty eyeballs, stoned out of my mind on esteem, hungry for days spent in the comfort of the office making lists and ticking things off. And I was so happy. The meaning was back, the sense of purpose.

Suddenly we had a hundred and one things to do. We would need to go back to England to get renewed visas (I would need a work visa and our plan to renew tourist visas in Nepal was now rendered defunct), pack up and leave Goa, return our car and store our furniture and electrical goods for our return in six months. The money offer was good and the company had agreed to pay return tickets for us, so the endurance test that was Syrian Arab Airlines could be avoided in favour of the luxury of BA.

****

In the midst of our leaving preparations, our friends Tom and Emma arrived for a two-week holiday, and in celebration we elected to take a holiday with them. They had taken the same deal and ended up at the same despicable hotel as Lucille and Jim, so to avoid all hassle we had them stay at the Casablanca to acclimatise, after which we set off together for a few days in the beach huts which stand proudly constructed on coconut palm stilts on Palolem beach. We drove the 50 or so kilometres as a farewell trip for our beloved car. Our companions were handling Goa pretty well and had neither scratched their Gucci sunglasses nor dirtied their Patrick Cox, but were visibly windswept and slightly shaken by the bumpiness and occasional swerving insanity of the trip. The fact that I had hit a defiant cow on the road, leaving it with a sore head and a bruised ego was quickly overlooked when they saw the magnificent arc of Goa's coolest beach for the first time. We checked in to one of the many beach huts and commenced the business of consuming holiday fluids. That night I got the most drunk I have been for many years. I think the combination of cumulative tension brought on by such a rapid and unprecedented career path decision and the confusing sense of both failure and success that had settled upon my conscience as a result conspired and ordered me to get unconscious. I awoke with only one recollection of the night, which was that we had jointly vowed to find and excrete in a pig toilet.

Dry toilets where pigs are kept to clean up openly dumped effluence used to be very common in Goa but the vanity and sanity of the tourist comes first and they are now all but departed. All but a few that is. I had spotted one of the few on a stumbling excursion in search of a suitable spot for a leak. On my return I explained the concept to my disgusted friends and by all accounts talked of nothing else for the rest of the night. After a great deal of persuasion consisting predominantly of taunting "Chicken, Chicken, Chicken" I secured an agreement from Tom to accompany me on my mission. The next morning he was unconvinced, saying that he would instead have a civilised breakfast and afterwards would spend time on the beach with the beautiful people, Palolem's seasonal population of young Europeans and Israelis who spend the day on the beach looking tanned, lithe and nubile and smoking huge tank slapping reefers and monstrous chillums, spending the evening in the bars doing pretty much the same thing. But I was determined and set off with only my digital camera as companion and recorder. To find the toilets, I simply followed the pigs, quickly coming upon a small hut made from coconut matting with two hardened mud blocks on the inside and a little ramp or shoot between them pointing outside. This was clearly not a public toilet but was owned by local people. Fortunately for me it was not locked, as having to ask to use someone's pig toilet would need a good deal more pluck than I could muster, particularly with such an uncompromising hangover. Instead I waited around until the coast was clear and surreptitiously nipped inside. As soon as I was inside the pigs gathered around the back and within seconds there were excited grunting noises and a number of pink and black snouts poking expectantly up through the shoot. I would have to work fast as the rising noise level might attract attention and I didn't fancy getting caught on the way out, "White Man Trespasses in Pig Toilet " - explain your way out of that. Getting into position, the noise and jostling of the pigs was getting louder and more animated, I began to feel the full strength of my hangover as the ultimate strangeness of my circumstances became more and more apparent, and now stage fright kicked in as I squatted with my shorts around my ankles and the pigs squealed and bumped up against the little hut that was feeling less and less stable by the second. Finally I managed to impart my intensions and became immediately and horribly aware that the pigs were in approval, with very little to eat the night before I had expected what a doctor might term 'loose stools' but Niagara Falls I did not anticipate, and clearly neither did the pigs. But they weren't complaining - so much so quickly and so indiscriminately meant there was plenty to go round. As I threw water down the shoot and skulked away, snapping the pigs from both inside and out, I noticed a little boy watching me and scuttled off fast lest he should call his dad. After a safe distance I perused the snaps, it was all there to see. With arms aloft I walked back to the beach in triumph. As I got back to the beach huts my friends were still having breakfast in a shack and I sat down smugly next to them.

"Mission accomplished. The geese fly south for winter. The eagle has landed, and I have pictures to prove it."

As I readied the camera to prove my manliness the waiter approached me.

"Breakfast sir? Bacon? Sausages?"

"Err...maybe just a cup of tea."

Chapter Eleven

It was time to say goodbye to Goa, at least for the time being. Packing up our stuff less than six months after arrival seemed like cheating somehow. It felt as though we were copping out on our intentions, but we would be back to India in two weeks to stay for another year and I kept this in mind while packing. I was certainly glad we'd had the foresight to insert a six-month break clause into the rental contract. Being able to return some of the unnecessary junk we had hauled all the way out here also felt somehow fraudulent, and it was with a hint of embarrassment that we threw so many unused items back into the suitcases to return home for good. We opted to dispatch the belongings we would need to keep in Pune by way of a courier, enabling us to return with suitcases bulging with Twiglets and very little else. In an effort to find a suitable receptacle for our mounting accumulations (homewares mainly), we set off to Mapusa.

We had two days left before leaving and driving the excellent little car that had been so surprisingly reliable into the scruffy town for the last time was tinged with sadness. Unlike our friend Max who had inexplicably fallen in love with Mapusa's filthy, smelly, stifling atmosphere, I loathe the place. I hate it for its uninterrupted street hassle and the unmitigated treachery of navigating its congested roads and infested streets. I hate it to the point where I had been singing, "I don't want to go to Mapusa" in the style of 'I don't want to go to Chelsea' for the past three months. But today seemed special and something stirred within my sensibilities, telling me to make the most of things, so it was with a sense of merriment and tomfoolery that we parked up the car and embarked upon our final Goan errands. We found the courier recommended by a friend and informed them that we would send someone with our parcel later that day asking what type of box or container would be most appropriate. "A strong one" came the obvious reply. We had seen some 'monsoon chests' in the market, large blanket boxes fashioned from lightweight tin and I asked whether these would be ok. I received the head wobble, which I took for a yes, and we headed into the market to find one. As always the hassle started instantaneously - "Hallo Sir, look at my shop, looking is free, Lovely Jubbily, cheaper than Asda" (the misappropriated rhyming slang and allusions to Englishness were the most grating as always, I try not to get insulted, I try you understand, honest, I really try). As we walked through the fruit and vegetable sections, the usual "Hallo Pineapple?" "Hallo Orange?", rang out and the retort of "No I am not an orange" could not be resisted. "Hallo seeds?" offered by a man selling heaps of pulses enabled "we are no longer seeds sir, you will notice that we have germinated into fully grown human beings" and Julie's groans were heard once again.

Suddenly I saw a man selling sheep, or sheep-goats as their ambiguity dictates we must call them. These animals are readily available alive or dead in the market and a sight I had seen many times before, but something was different, these beasts were being wielded aloft by one leg and were rising effortlessly above the crowd at the hand of their vendor, who was in turn crying "ship, ship, ship, ship, ship" at the top of his voice. As I approached him I realised they were stuffed, and as I drew nearer I saw how fantastically stuffed they had been. The legs had been extended with internal wooden splints, all orifices had been stitched, the mouth had disappeared completely, the eyes had been replaced with buttons and the ears had been supplemented with curly horns fashioned from wire, masking tape and marker pen. There were three available and I of course wanted the lot. The vendor seemed surprised at my enthusiasm, it seemed he was getting only rather disgusted expressions from the other tourists, and his starting price was low "two hundred rupees only". I conferred with Julie as we marvelled at the three inert offerings deciding how we would cope with buying all three. The vendor starting getting nervous, "One hundred fifty rupees only?" I tried to explain our indecision, "It is not a matter of price, two hundred is excellent, it is a matter of how many and how we get them home." The vendor remained confused, "OK, one hundred rupees." I took money from my pocket to reassure him, "Here's 200 rupees for this one" I said handing him the cash and taking the tallest, most misshapen example. I turned to confer with Julie:

"How can we get them home? Maybe we could ship them? No, I can't see them getting through customs intact, imagine the contents label, 'three stuffed sheep'. What about taking them on the flight? No, maybe not. Well, we'd better only buy one then. What are we going to do with this one anyway? Take it to Pune with us I guess."

Having answered all my own questions I sent the vendor away with some regret (by the time he had finished knocking himself down we could have had the other two for fifty rupees each) and set off into the crowd. A man carrying a dead animal is fairly commonplace in the market, particularly in the butcher's row, but a white man sporting a stuffed animal of doubtful pedigree with its stiff extended legs straddling his neck is apparently very uncommon. I had no idea one could cause so much alarm. And for the first time I felt a change in attitude from the stallholders. In an instant my position had changed from potential customer who must be harassed into parting with money by any means possible, to state clown. I had no idea carrying a stuffed animal could be so funny. In truth I have never seen so many Indians laugh so much before or since. Suddenly I was everyone's *friend*.

I was accosted by at least every fifth person I passed who wanted to stroke or poke my dead shoulder-pet and ask me various things about it. The main avenues of enquiry were based around *why* I was carrying such a thing, *what* it was actually for, and of course, *how much* I had paid for it. To each question in turn I replied:

"I am carrying it because it is dead and unable to walk",

"It is for good luck" and

"I'm sorry, but it is of no consequence how much it is, it is not for sale. This animal is mine."

One man, a very serious sort who seemed overly perturbed by the spectacle, remained adamant. As he questioned me we stood by a stall selling gaudy religious artefacts and amulets for good luck.

"But what is this for, it makes no sense?" He put forward, his smile waning.

"You mean to say that you look at these items and you make better sense?" I retorted motioning toward the stall with greater intent to perplex.

"Now you are making no sense. Tell me, what is this animal for? What are you doing with it? Why do you have it?" He persisted.

"I have it only because I love it. Good day sir." And I walked on.

Causing a storm had not been my original intention and remembering the business in hand I returned my attentions to finding a 'storm chest' while the tumult raged around me. Upon finding a suitable box, an enormous silver trunk, which despite its utility nature was quite beautiful, I paid its bearer and began to cart it off back to the car. Before I had taken two full steps I was confronted by two *coolie* boys, one around ten years old, the other not a day over six. They insisted on taking my new possessions back to the car for cash. I abhor the notion of child labour, and always feel very uncomfortable when confronted with kids I'm supposed to pay. In India this contention can be frequent, particularly where tourists are prevalent.

In a country with no welfare state but compulsory schooling it is very easy to say 'you should be in school' to a kid who has been sent to scratch around on the street for a few pennies by desperate parents. I once bit the head off a beach shack owner for letting a ten-year-old child serve the lunch I had ordered, saying "it's a disgrace, this child has a right to an education, you shouldn't be exploiting him in this way". Once I'd calmed down, the shack owner told me very evenly that the boy in question was deaf and no school was available to him apart from a very expensive out of state option. He also told me the child had been abandoned by his parents and was getting an education from the shack owners at home in the evening. He also reminded me that it was Saturday afternoon and that even if there were a school to take him he would not be in attendance right now. As I sunk lower and lower in my chair, he told me finally that over the past ten years he had successfully persuaded numerous foreigners to fund the education of the boys and girls he introduced them to in the safety of his employ. He told me with some pride and no malice that he had been responsible for the education of no less than ten needy kids himself and held out great hope for the boy in question as an elderly couple from Birmingham were in the process of discussing terms for a ten-year school programme with the authorities. Nothing is ever black and white in India, something we learn very quickly.

So confronted with the boys I gave in. The elder boy took the lightweight but enormous chest (easily big enough for him to sleep in) and the younger boy, with some glee, took the sheep-goat, which he immediately put on his head and set off in front of the boy with the huge trunk who in turn led in front of Julie and I. The insane procession led through the multitude causing less commotion than when I had been wielding the stuffed animal but still creating quite a fuss nonetheless. At one point as we turned a tight corner, the small sheep-goat headed boy bumped into a flighty woman who saw the animal and screamed in terror, slipping sideways into the road. Upon righting herself she clouted the boy with an extended open palm, sending the animal scuttling along the floor, and shouting "Devil Child" ran straight into me. "Get a grip on yourself my dear woman" I opined and shimmied left to commence the preposterous parade. We reached the car and paid the boys a handsome 50 rupees each, at which they both looked momentarily serious, bowed a little and sped off screaming. I can only assume they were happy with their compensation. For such a round the bend opportunity I considered it money well spent.

We arrived back in Candolim, packed up the trunk, sheep-goat and all and called Peter to collect it and take it to the courier. As he returned to us with the docket, we arranged for him to help us put our white goods into storage the next day at the Casablanca Hotel, where Gurmeet had very kindly offered to keep them dry during monsoon (I knew damn well he would rent the fridge and the TV out to his guests until he closed in June, but what the hell? Let him make some more money out of us, everyone else does). That evening we went out to say a temporary goodbye to all our friends and as I looked around the bars and restaurants I felt it was indeed time to leave. I realised I was ready for something new, something different, and I felt optimistic it might be found in Pune. While Goa is a very beautiful place, it has its drawbacks. Candolim is really the Bournemouth or Bridlington of the East, a place with plenty of amenities and amusements where old people can come, sit on a beautiful beach and think of England. Cutting edge culture it is not.

The next morning Peter arrived with a goods rickshaw and driver. Onto this tiny, waspish machine was loaded our gear and the pair set off, the fridge swaying dangerously as the three wheeler negotiated the inevitable bumps. One great thing about India is that hired help comes so cheap; the move cost us less than a pound. Moments later Raj arrived to collect the best car I have ever had. It was heartbreaking to watch him drive the little vehicle away, and I vowed there and then that there would be more to this story, we'd buy one in Pune, no doubt about it. As we waited for Peter to come back in his taxi to take us to the airport, Arnold (who we had tried hard to take leave of the night before), Isabel and their kids came to wait with us. As Peter arrived and our casket sized cases were loaded up it was an emotional Arnold who gave us leaving gifts of cashew nuts and assured us that we had been "his first and his best guests" and that we "had trusted him and had been honourable throughout". As we sped away from the strange orange house the family waved from the verandah and I considered how many other landlords had come to see me off with gifts and sincere wishes in my life. India is so different. Everything is different. Nothing is the same.

As we arrived in London after a seamless twelve-hour journey, I vowed to tear up the Syrian Arab Airlines ticket I had in my wallet. It would be British Airways all the way from now on. As I joked happily with Julie I attempted to take the tickets out of my wallet to tear up, only to find I had left the very same wallet on the plane. The wallet containing cash, credit cards, tickets and above all, the docket for the parcel we had shipped to Pune to be there at our arrival.

I spent the London mosquito break drinking dark beer and phoning Sarasvat and the courier company to see if my rather expensive next-day delivery had reached its destination. It hadn't. After ten days it still hadn't and the courier company started to deny it ever existed – "without ticket, no proof, we not do anything". I could not get the stuffed sheep off my mind, but eventually resigned to never seeing it again, reassuring myself that I would be able to buy another back in Goa. The following day Sarasvat called to say that a rather battered looking case had arrived and that while both locks were broken and he could not vouch for everything that should have been in it, it did indeed contain a stuffed sheep which was intact and currently gracing his mantlepiece much to his wife's displeasure. Worry over; I concentrated on the dark ales.

Having secured a one year work visa and an accompanying document enabling Julie to stay with me, we said goodbye to family and friends and boarded the plane back to India. It had been an unusual trip. Neither of us had been ready to come home and it somehow felt *wrong*. We wondered how we would feel in a year's time. Would we ever want to go back?

****

Upon arrival in Mumbai my new corporate life began immediately. We had to report to a hotel on Juhu beach for a three day company 'offsite' where I would meet the senior management for the first time and we could presumably play corporate games, plan the coming year and 'bond' together. I was no stranger to this type of event and was looking forward to seeing what the rest of the staff were all about. The hotel was a dowdy business lodge positioned just downwind of the fish-drying beach north of Mumbai. It had not been modernised since the mid 1980's when its four star status meant something. Some fifteen plus years later and its rating had slipped past three stars. The airless rooms were damp with peeling paintwork and a musty smell indicating that the air conditioning had not been used for some time. This was not the luxurious start to my new life of money that I had expected and I made this very clear to my boss on meeting him first thing next morning.

"Ah Avinash, how are you? Who's responsible for us staying in this shithole? "

In retrospect, this was probably not the start he was looking for from me either.

I met the rest of the management and sales departments for the entire group (it's a huge company) in the first meeting that morning and listened intently as everyone introduced themselves. There were four Americans around the table, one Australian, one Brit (me) and twenty-five Indians. When my turn came I opined about how I had been working with the company for a few months already and had a good idea about what needed to be done to improve things. I made it clear that the problem areas were communication with clients (not being able to understand their Indian colleagues) and an overall fear of clients by junior Indian staff who were inexperienced in being able to say no or 'manage expectations' to use the commercial terminology. I looked for recognition among my new contemporaries but was shot piercing black looks all round. I had obviously jumped in far too early with my criticism. The group chief executive looked at me with visible disgust and urged the next speaker to get on with it. It became clear to me there and then that my opinions were not of interest, particularly where they involved criticism of Indian working practices.

Later that day the conversation turned to finances and I again blurted imperiously on how things would have to change if the business (or at least my part of it) was to succeed. I told them directly not to expect a profit in the first year of (my department's) trading and informed them that the nature of my part of the business was a low profit margin affair and that even with the vastly reduced costs in India, the company should expect much lower margins than in other business areas. I really should have kept my mouth shut. This time another chief executive (everyone in India is a CEO of something, you can probably become the CEO of tea and biscuits at this company) turned on me.

"How do you know this? Do you know anything at all about the Indian BPO market?" he snorted disdainfully.

"Yes I do, do you know anything at all about market research? No, I thought not."

And right there I had burnt any possible bridges to future advancement. There is no room for argument in big companies, there is only sycophancy. I reflected upon how many times I had forgotten this over the years, and concluded that whatever the number, I would never ever be asked on to the golf course.

That evening there was a 'pool party' where other staff members joined us for drinks by the pool. The swimming pool party is a common occurrence in Indian business hotels but is much less exciting than it sounds. These affairs comprise largely of fully clothed men standing somewhere near the hotel swimming pool eating a bland Indian buffet quickly and silently and then leaving in a flurry of handshakes. Within an hour, everyone worth talking to will have left and the three most boring, unmarried men will remain. These tedious bores will then ask you how long you have been in India, how you like it, how long you intend to stay and how much you pay in rent for your house. This occasion was slightly different however, my friends were coming and drink would be taken. While waiting for them to arrive, Julie and I convened at the pool at I introduced her to those people whose names I could remember. While speaking to two of the American sales people, whose responsibility it was to bring the work into the company, it became very clear that neither were even remotely enthusiastic about India. This was the first trip for both of them and their distaste was evident on their faces. They asked us questions like:

"How can you live here? Its filthy and it stinks."

"What do you eat? Surely you don't eat Indian food?"

"How do you cope with all the poverty? Really there are so many beggars, it's horrible."

I gave them short shrift, reminding them that their livelihood rested upon India and they'd better start liking it. I asked them how they expected to sell business to US clients if they themselves were unenthusiastic about it. They told me the clients didn't care about that, they just wanted to save money. Julie and I walked away with our heads in our hands.

We got a very different response from my new Australian colleague who was in a similar situation to ourselves and had been in India for some time before being recruited and was very enthusiastic about the place. She said she assumed we had been here a long time because she noticed me drinking the water at the table that was quite patently not mineral water. I showed off that I had got used to it and had become completely accustomed to Indian food as my insides churned at the realisation that I had had three glasses of the water that I had just assumed was bottled. Well, at least I was now prepared for what tomorrow might bring.

When my workmates finally arrived after a two hour detour where their driver got lost, the group chief executive, now *dressed down* in horribly tapered snow-wash jeans and a company-logo polo shirt, addressed the assembly with a wholly uninspiring and unconvincing speech about what a great bunch they all are and how integral they would be to the success of the company. Julie noted that he was clearly not addressing me. The gathering whooped and clapped in the American way despite the flatness of the delivery. I asked Sarasvat and Shanti what they thought of the chief exec and they both replied that they 'liked him'. I asked why and they weren't able to answer to my satisfaction, I pushed them for adjectives to describe him, shooing the inevitable 'nice' and 'good' away like pesky flies (Indians use few adjectives in English, the word 'nice' is applied to pretty much everything), and they could only come up with 'competent', 'accomplished' and 'does a good job'. I put it to them that he was a 'flatly aloof bore with an attitude of ill conceived superiority and a complete absence of social skills' and that he was also unequivocally a bore. Shanti reminded me that she wanted to 'get on' in the company and I left it at that.

After a few drinks the singing started. A guitar was produced and a small plastic table was converted into a *tabla*. After some very spirited renditions of Hindi film classics and of course 'Take Me Home Country Roads' and 'Hotel California', I sang a song and urged my foreign compatriots to follow me up afterwards to do the same. There were no takers. Just like everyone else, I loathe the situation where people are squirming behind one another trying to avoid taking part in an amateur sing song. But being brought up in an Irish family, this was a commonplace part of my youth, and there is an expectation that the most outgoing members of the group will all make some contribution, even if it is playing a penny whistle with one hand while throwing darts with the other (my uncle's unfaltering input). In this spirit I picked upon the senior management in the group and asked directly for their party piece. I was amazed to see them cower in embarrassment while the lower management were in fact itching to get up and perform, the *big swinging dicks* among us were as I put it 'too chicken shit to take part'. This left a bad taste in the mouth for me and while the evening raged on in song and I saw them creeping off to bed, I formed opinions about these people that really shaped my whole attitude to the new job. In short, I thought of them as tapped out and spineless, not the sort of person one is looking for as a *boss*.

Arriving in Pune a couple of days later, Julie and I put up at the Central Park Hotel, a slightly cheaper hostelry than we were used to, but a first rate business hotel with rooftop pool nonetheless. And of course, a room with a bath. While I got to grips with my first couple of weeks of full time work, Julie set about the arduous task of finding us a place to live.

Being some seven or eight kilometres from the office, and given my stature (AVP - Assistant Vice Principal apparently) I could command the use of an air conditioned company vehicle and driver to take me to and from the office. In the first two weeks the driver was either early or late every day. When they were early I would get hassled by reception to hurry down to them, when they were late I would call the office to be told "driver is coming sir" and have the phone put down. Once I called angrily to speak to Sharmila my long-suffering reception contact to be told 'Sharmila is having cup of tea sir' after which the phone was slammed down again. The problem was twofold: one, the drivers were either early or late, and two, the receptionists could speak English but not understand it if it was spoken by an Englishman - each time I called, they tried to converse and then freaked out and put the phone down on me. I brought this up with the head of administration whose 'I just can't get the staff' attitude was to become a particular rankle for me over time. Another enemy made.

When I was picked up, I was being driven around by Mitul, the slowest, most careful driver in the country, frustratingly careful to the point of being dangerous. Unlike all other taxi drivers in India, and very much like all taxi drivers in the rest of the world, Mitul considered everyone else on the road a 'madman'.

"He is madman, no signals, no horn, madman, very dangerous."

We crossed the Pune flyover one morning after he had collected me from the Central Park Hotel. There had been a minor accident or incident on the road, a small three wheel goods carrier was stopped, nose in to the roadside at a 110 degree angle, two cycles were also stopped, one of which was being examined by two men, one of whom was presumably the owner. Between the cycles and the three-wheeler on the road was what looked like a skinned circular dead animal. We passed slowly, feasting our eyes first on the repugnant pile, then on the men involved, who stared back deliberately into our eyes. Our motion seemed to slow, the tenet of our fixed gaze holding us motionless for a second. The scene reminded me of Fargo, just after the first murder, when the witnesses drive by and see Steve Buscemi holding the body.

"What was that?" I asked Mitul, disturbed by the scene, unwittingly considering what the mass of what looked like a decanted bucket of liver could be.

"You know Non-Veg?" the reply.

****

The joining formalities were exhausting. It took me fully two weeks before I could be issued a staff pass to enter the building alone. This entailed presenting myself to the Commissioner of Police - Registry of Foreigners Department where I waited four hours for a signature on a form that I left on the wall outside after waiting there an hour for my driver to show up. I had to open a bank account, take insurance, a pension scheme, register for tax and a multitude of other formalities represented by a multitude of other forms, and all in triplicate. Although I had already got the job, I had to fill in an application form that asked me to fill in some blanks: 'My colleagues think of me as…' 'My boss thinks of me as…' and 'When I'm frustrated I…' I will not divulge my answers, but I will divulge that the form came back to me via the weary administrator to fill in again with the word 'spoiled' written across it in angry red ink. Somehow I had managed to get away without having a medical, but this left me with the problem of identifying my blood group, which until this episode, I had no notion of. I told the administrator that I had no idea of my blood group and he looked at me as if I was insane.

"But what happens if you have an accident and you need blood?" came his incredulous response.

"Well I hope my blood would be tested and a corresponding type used. If I were unconscious I wouldn't be able to tell the hospital anyway. I assume they wouldn't just start pumping me full of blood on my say so?"

He gave me the unsure head wobble, saying, "Well, I have to know, how will you find out?"

I thought for a second, "I know, I'll phone my father and ask him, he'll know." As I dialled a fictional number and listened to the operator saying 'please replace the handset…' I reflected that there would be more chance of the Queen knowing my blood group than my father.

"Hello, Dad? Hello, its Mick, what blood group am I? O? Thanks bye."

"I am blood group O."

He smiled a delighted smile, "Same as me."

(It was a lucky guess too; I checked when I finally did take a medical on the insistence of the Über Administrator, or 'The Tyrant of the Tape' as he is universally known).

When I finally got my access all areas pass, I was free to use the employee's entrance and thus, the more scruffy internal lifts. Based on my experience of the same I made up a joke:

Q - How many Indians can you get in a lift?
A - Eighteen.

I met an Englishman who I had been told was in a similar situation to me, living in Pune and working full time for the company. I had expected him to at least be up for a bit of colloquial chat, perhaps even someone with whom I might become friends, marooned as we both were on an island of Indianness. I could not have been more wrong. He looked aghast when he saw me with my employee pass round my neck, making me feel like a gatecrasher at the Dorchester Club, shunned as riff-raff by the affiliated. I spoke to him for a while and allowed him to tell me in a snooty Middle English manner how his kennel bills back home were killing him (darling) but how he hated England and everyone in it, had no intention of ever going back and had no desire for anything English at all, least of all football. With this he flounced off with a palm raised at the sun, the back of his fingertips touching his delicate brow. I stood alone and looked myself up and down, wondering if I looked like a football fan or some kind of intolerable English yob, perhaps he just didn't like Mods. It was not until the second time I met him that it dawned on me that he might have been concerned I would blow his cover - he was presumably gay, or at least very effeminate, something that his workmates might not pick up on here in India, and possibly something he had become tired of dealing with in England. Either this or he was just plain rude. Either way I left to him to it, proudly concealing my disappointment.

Getting a cup of tea was proving even harder than getting in the lift. With some crazy man-of-the-people notion, I had said it would be unnecessary to have my own office (or 'caybin' as they're called here, I think it is supposed to be 'cabin', but surely even Americans don't say 'caybin', do they?) and that I would be quite happy to have a desk on the open floor. I quickly found out that this prevented me from eating or drinking at my desk as a stern, matronly looking supervisor told me with some vehemence. Getting a palatable cup of tea meant bringing a flask. Tea without sugar is looked upon with raised eyebrows in India, and as this is my preferred cocktail, I resorted to the BYO approach.

I could tell from the frustrations of my early interactions with the great cogs of the company wheel that I would have to be forceful and outspoken if I wanted to make a success of my time here. Not only was I delayed by drivers, so were all the staff. Not only was I a victim of the unresponsive but 'jobsworth' IT department, everyone suffered. And it was not just me who felt the frustrations of the slow grinding cogs of administration and the pedantic signing and countersigning of the finance department, the whole company felt it. I asked my colleagues whether the problems I was facing were particular to the company or indicative of what I could expect throughout India. The reply came in the form of a huge accusatory finger pointing menacingly at the company logo hanging from the ceiling in the middle of the office floor. Sarasvat told me "At my last company if you needed anything you just sent an email to the relevant person and expected a reply within an hour. If they can't fix your problem they will tell you why immediately. If your driver is late and you lodge a complaint, you can expect an apology from the head of admin along with the reason he was late within a couple of hours. This is for agents, managers get an immediate response. Managers at this company get no more respect than the cleaners."

Indeed, the plight of managers is rarely considered when contemplating Indian call centres. The assumption is always that the staff who work on the phones have a terribly tough time dealing with all the irate customers, and that the supervisors are heartless bastards who whip them and make them work without breaks for hours on end for little reward. This may well be the case in Europe and the US, but in India, the supervisory staff are predominantly very employee focused (as the companies also tend to be) and are often at the mercy of fickle employees who will get extensively trained on a process, do it for a couple of days, decide they don't like it and move on to another call centre that has just opened around the corner for more money. Staff attrition is a huge problem, in Mumbai and Gurgaon staff attrition rates are expected to be as high as 60% per annum for all processes. The assumption in the west has always been that this is because the jobs are terrible. The truth is that somewhere around 90% of those that leave a call centre job do so to join a competitor call centre in a very similar position. The training involved makes attrition at this level very expensive, and the supervisory staff are paid in accordance with this. In some cases up to 25% of their expected salary will be deducted if the attrition level on their process is in excess of the stated average. This means that supervisors will do anything to hang on to their staff, begging and pleading with them to stay, even offering them money from their own pockets on occasion. This role reversal can be hard to take for those who have chosen a career in the call centres, taking on responsibility is rarely the position of power that would be expected and often brings far more stress and anxiety than is bearable. The percentage of those leaving the call centre industry after a couple of years at management level even exceeds the percentage leaving without management experience.

But none of this hassle was going to get to me. I was determined to deal with whatever the company threw at me and guide this team to a very comfortable and hopefully lucrative position. If the driver was late, I could deal with it. If meetings didn't start on time, I wouldn't let it bother me. But if I didn't get out of the hotel soon into a place I could call home, I would go insane.

The Indian business hotel is an air-conditioned comfort monster, spewing out extravagance and exasperation in equal measure. Seemingly a polar opposite to the outside world, this institutionalised corporate cocoon is in fact a microcosm of the Indian Experience - it is both exhilarating and frustrating at the same time. Each day is the same, it is only our response that changes. One day's aggravation is the next day's amusement. Everything remains constant, it is simply our mood that alters it. The problem is this swinging mood. At home one is able to stabilise this by doing whatever one pleases, where as in a hotel, particularly in an Indian hotel, one is at the mercy of the efforts of the staff to do everything their way and in the time they want to do it. And there are so many decisions, selections that are normally enjoyable and simple become horribly complicated after too long in a hotel. Painful decisions such as whether to take the free but terribly frustrating breakfast (see further gripes below), whether to dine in the room and spend the whole evening in a reverie of unselective, non discerning TV watching or to brave the dining room for an hour of slow, confusing, bowing and scraping waiters bringing the wrong order and apologizing for things they haven't done wrong. The irritation at having forgotten to put the Do Not Disturb sign on the door and the resultant 8am "Ding Dong - Hallo Housekeeping". Then there's being watched while exercising alone in the gym by the bored but eager to please gym attendant and having to ask him to leave the changing room to prevent him from watching you change, and trying to avoid people thrashing about in the pool, swimming in a fashion but never submerging their heads, and with no notion that you are attempting to swim in a straight line (can Indians swim? I've not yet met one who can). And of course there is the slowness of time to deal with; spending far too long in the bathroom using all the products from sheer boredom, having a shower after having a bath just because you have time, and waiting for the digital clock to reach a suitable number such as 12:12 or 12:34 before doing whatever it is you have decided you are going to kill time with next.

The problems in communication are an added frustration. In the midst of too many decisions confusion reigns anyway, but not being able to make things immediately clear leads to further frustration. For example, at the Sun 'n' Sand Hotel coffee shop, Pune, I went through the following conversation while trying to order a quick lunch before heading off to work (I would start at 2pm IST, 9am GMT).

"I'll take the chicken tikka sandwich please."
"Grill."
"I'm sorry?"
"Grill, grill, grill. Grill is more better."
"What on earth are you talking about?"
"Grill is much more better than plain or toasted sir."

I looked at the top of the sandwich menu card: 'Sandwiches - choice of any one plain/toasted/grilled.'

"Oh, grilled please."
"Yes, grill is more better sir."

And at Central Park Hotel, the following fiasco:

To the waiter, "I have only half an hour to order, be served and eat. Do you think you can serve me quickly enough? Can you get me in and out in 30 minutes?"

"What do you want to order sir?"

"I don't know, how about we start by you giving me the menu and me asking you the same question again but with direct reference to my chosen dish?"

"Which dish sir would you like?"

"Give me the menu. The menu."

After some perusal, "OK, can you make and serve me the Greek salad in 30 minutes? I must leave in 30 minutes."

The head wobble.

"Can I take that as a yes?"

The head wobble again.

"Yes?" I shout, nodding furiously.

"Yes sir?"

Oh my God another question. I began to show frustration. I turned to Julie, who was fixing me with a knowing smile.

"What are you going to have?"

"I'll have the cream of coriander soup, nothing else."

After the last few months Julie knew not to complicate the ordering process, I tried to take my lead from her excellent example, normally I failed, but this time I really was in a hurry.

"Right. One Greek salad for me, one cream of coriander soup for her. One bottle of mineral water. That is all, thank you." I handed him back the menu. "Hurry now please, I only have 27 minutes before a car arrives to take me to the office."

"OK sir."

I turned to Julie, "What are the chances?"

I knew damn well what the answer to that was, we had been eating breakfast and lunch in this hotel restaurant for the last three days, and the fiasco of getting eggs *on* toast which had first been amusing had become tedious; my knuckles were starting to whiten.

The first morning's attempt to get poached eggs on toast had resulted first in cold toast being brought immediately, which I of course rejected, followed by two poached eggs on a plate, with of course no toast with which to eat them. I saw the error and corrected my ordering technique the next morning, stressing very clearly and with some 'you remember yesterday?' type irritation, "Two poached eggs and two toast, AT THE SAME TIME." Still no luck, this time the toast again came out immediately which I quickly rejected saying "bring it AT THE SAME TIME AS THE EGGS please." The same toast was brought out again after some time, by now very cold, and on a different plate to the eggs. I gave in and ate. The next morning I was determined. I felt that the miscommunication was largely a result of me speaking English like an Englishman and not like an Indian. I attempted to resolve the issue by first asking for a pen and the waiter's order pad, I wrote "2 poached eggs and 2 toast ON THE SAME PLATE, AT THE SAME TIME."

It's never as simple as you think, the confused waiter brought out the manager and I had to explain how I simply wanted what was written on the pad and that if they read what I had written instead of trying to ascertain why I had written it, we might get somewhere. Eventually I got my order, the eggs were hard and the toast was cold. While eating I got that cold shiver which reminds us that the chef can do whatever he likes to our food while cooking it, and that it is never a good idea to kick up a stink *before* ordering a meal.

"You know exactly what the chances are" she replied with amusement. Unlike me, she was in no hurry and was quite looking forward to the inevitable saga unfolding.

Five minutes later the Greek salad arrived. I tucked in my chin and glanced up gloatingly at Julie. It was good too. Five minutes later I had finished it and the hovering waiter approached me for the plate. Normally I would wait for others at my table to be served before starting my meal, but I had mitigating circumstances. In my defence, I had enquired about the arrival of Julie's soup once while eating. The waiter reached to take my plate.

"Where's the soup?" I barked.

"Soup is coming sir" he replied, desperately trying to remove my plate despite my flailing arms.

"Is it the policy of this restaurant to serve diners one at a time?" I was smiling, lulling the man into false security.

"Sorry sir?"

"What are you sorry about? You're always sorry, are you sorry that the soup hasn't arrived? Or are you just sorry like everybody in this place is sorry?"

"You want a cup of tea sir?"

(He can't have said that, surely?)

"What?"

"You said tea sir? You want a cup of tea yes?"

My anger dissolved. I had been disarmed by one miscommunication too many. I laughed aloud, unable to resist.

"Yes. I want tea, one cup of tea with no sugar and cold milk. Make it a good one as well, a really good cup of tea."

"Old milk sir?" he was looking terribly perplexed, perhaps ready to cry.

I rose, "I have to go now, thank you so much for the experience. Enjoy your soup my dear." I kissed Julie and left.

"See you at breakfast." I said to the waiter, wobbling my head at him and laughing like a drain.

Chapter Twelve

Julie had spent almost two weeks doing little more than waiting for estate agents to turn up and occasionally viewing the odd flat. The system of finding a rental property in Pune is less developed than in European cities. Unlike Goa there are some official estate agents, but many places are still only available through word of mouth. Certainly, few landlords are prepared to pay an estate agent to find them a tenant; it is therefore largely down to the tenant to pay the agent. Being white and having a business card meant we were top priority upon stepping into any of the four or five official estate agents in Pune, in fact we should have been able to see ten or so flats and make a decision within two or three days, but things are rarely that easy. All the reasonable properties in the city are built privately within security patrolled compounds called housing colonies. These usually comprise a number of high-rise blocks and a couple of rows of bungalows surrounding some gardens and possibly a swimming pool. These colonies are currently mushrooming all over Pune to meet the increasing demand for middle class homes as industry expands in the city and the economy booms.

The trend among the wealthy is to keep buying new properties. The value of the quite beautiful Art Deco blocks and bungalows around the Boat Club Road area is significantly less than a boring new-build on an estate in the east of the city, it seems period architecture is synonymous with tradition (albeit often European) and not progress, which today's homebuyer wants. This trend means that older properties, maybe six or seven years old are left to stand empty while new properties are eagerly sought. Given that a general attitude of latency towards cosmetic repair and maintenance and an utterly ambivalent approach to aesthetic finishing is so well established throughout India, the housing colonies over three years old look something like Eastern European housing estates circa 1985 - dirty, once white/now grey, rendered concrete blocks of such a bland and functional nature as to defy description. The newer buildings conversely are anything but bland, indeed they could be considered blandish*ments* - encouraging statements to buy, buy, buy. If the race is on to erect the most colourful, high-tech, ersatz living facility in the country, Pune is at the vanguard. Recent developments have seen modernism go haywire as angular functionality is adorned with a hideous array of pastel facades, fake turrets, Moroccan domes and Gaudi mosaic, all of which are quickly tied together with appalling wiring, the cheapest available aluminium windows and randomly selected iron grilles which are presumably chosen by the residents with total commitment to irregularity.

As a self confessed Mod, India's lack of uniformity and love of slap happy decoration is very often too much a kaleidoscopic mismatch for my taste. I would suggest that Indian's don't design, they adorn. Everything is decorated; nothing is safe. A classic shaped vehicle such as the 'bowler hat' Ambassador is not enough, it must have a huge bonnet flag post, lace curtains, a squad of dashboard deities and a selection of lemons and chillies hanging from the rear bumper before it looks good. Similarly a front door is not a front door until it is completely clad in mock teak, bearing numerous family name plates of different materials, some poorly painted religious insignia and five or six outsized locks of different shapes and sizes and of a quality so dubious as to be rendered ineffective by a swift tug. Even nature is improved upon, trees are painted, rocks are painted, and cows are nothing like acceptable until they are completely red with blue horns. This jolly exuberance was cropping up in some shape or form in each property we viewed and I began to lose hope of finding a place with any sense of the simplicity, uniformity and functional formality that us Mods require.

That is until we found Hermes Heritage in Shastri Nagar with its utility Beirut-style, stark angular blocks and David Hockney's swimming pool. The flat we viewed was a huge glass fronted west-facing penthouse, seven stories up with an eighty square metre sun terrace. Inside were three en-suite bedrooms, a kitchen and an enormous living room, all newly painted in brilliant white. The 1980's black, white and glass furniture, leather sofa and all, reminded me of an old Boddingtons's beer advert where a tuxedo and ball gown couple strut about with erratic movements past a black swan and white grand piano, listening to 'cool' jazz and oozing sophistication while swigging pints of bitter and saying 'by 'eck you smell gorgeous tonight petal'. This was certainly the place for us. It had its drawbacks, the terrace balustrade was wobbly and the drop was sure death, and the lift played Theme from Love Story at shrill volume while the door was open - whilst this meant that people would probably shut the door behind them to stop it and thus the lift could be called to other floors, it also meant that if they didn't shut it the music would play on, and knowing the average Indian's tolerance to noise it could very well play on and on. It also meant that every guest would be announced by Theme From Love Story, I wasn't sure whether this was good or bad but was prepared to take a risk. But the shortcomings were outweighed by the benefits - it was close enough to the office to walk to work, the best supermarket in town was on the colony campus, stray dogs were energetically expelled from the walled surrounds and best of all, we could watch the sun rise in the east and set in the west. The entire flat was delimited by windows. Indeed, at the right time of year we would be able to simultaneously view a rising moon through one window and a setting sun through another. I really didn't think we'd find such a luxurious and fitting place. This was another unexpected departure from the imagined India, but still unquestionably and irrevocably Indian.

We agreed to take the flat and settled a moving in date for two days time. Ten days later we were still in the hotel going bananas. It was the usual bureaucracy, which we should have been used to by now, and should have been able to handle, but our continued proximity had begun to really get to us and we started to pick faults with each other. We were playing on the boredom, looking for excuses to moan further and generally dooming and glooming about our situation. In fact we became indecently self pitying, moaning about how dull life is in a four star hotel, dejectedly dragging ourselves to swim in the rooftop pool while the pariah kites swooped down from above, whinging about the food service, fridge temperature, iron temperature. Outside the window we could see people building an adjacent hotel, carrying sand and ballast eight stories up rickety wooden ladders, getting on with their hard life in 40 degree heat, working all day long and then sleeping on the ground where they had worked all day, laying on the rubble and rubbish of the job. We looked at them through the window of our air-conditioned chamber and just looked away again, even more bored. I have always maintained that if you want to experience a catalogue of human emotions in a short time, just go on holiday to India. But the sentiment of world-weariness and boredom was a new one on me and came as something of a surprise. The hotel was the culprit; after a few weeks holed up there I would look at the TV and just see nausea, the cherry veneer and brass fittings became a stifling manacle to creativity, bearing heavily on the soul, and the bowing and scraping of the staff grated the brain.

Finally we got out, and just in the nick of time. Moving into the flat was like landing on new soil, and the excitement of possibility returned.

After the aggravation of waiting for drivers every day, I was now able to walk to work. It was May, and by now so hot that I had to use an umbrella for shade on my journey. The walk was down a scruffy dirt track and I was an unusual spectacle, a white man with a briefcase and office attire, red umbrella, in 40-degree sunshine walking past street dwellers in tarpaulin covered huts staring in disbelief. This was classic India, where juxtapositions are a stark reality of life. As I picked my way carefully through the dust, trying to keep my black suede shoes clean (hardly a good choice for India I know, but once a Mod, always a Mod) and slowly enough to avoid breaking too much sweat, the people I passed would stop still and gape, the dogs would freeze as they found a foreign scent, the buffalo herders would halt in their tracks and the buffalo would momentarily run amok, even the jungle babblers and mynahs stopped their incessant chatter as I passed. For the first couple of days I enjoyed the novelty of the contrast: the laptop and the buffalos, the suede shoes and the open sewer, a rain umbrella against the piercing sun, but after a week I'd had enough, the abject poverty on my doorstep finally got to me and whilst I never felt any danger, I began to feel that I was in some way deriding these people parading past them with my wealth on my sleeve. In the end I had to start taking rickshaws just to ease my conscience.

Taking rickshaws in Pune, and indeed in any Indian city is not for the faint hearted. But it is not just the bumpy ride and kamikaze attitude of the drivers, it is the problem of communicating exactly where one wants to go that presents the biggest obstacle. In Goa, all taxi drivers speak enough English to enable basic communication, and the state is so small and ultimately rural that they will also have a very good idea of where everything is. The problem in the cities is twofold, the language and the fact that no one knows where anything is. In a city like Pune that has grown rapidly in recent years the problem is magnified as drivers try to get to grips with new areas and landmarks. Add to this the fact that many road names have been changed from their colonial vernacular in recent times and you have chaos. Rickshaw drivers have not spent years studying maps either, navigation is usually undertaken through the method of *asking someone else.*

But the language problem is one that I could fix, all I would need to do is learn a little Hindi and the hindrance would be overcome, or so I thought. I had tried to learn a little Konkani while in Goa but had reasoned that the complexity of learning a whole new alphabet, coupled with my own inability to speak any foreign language without acute embarrassment (I blame the schools) and the fact that the language is practically obsolete outside of Goa made this an unbeneficial exercise so I quickly dropped the idea in favour of watching more cricket. But Hindi? Here was a different prospect. Yes it was a new alphabet, and yes I'm terrible at languages, but it is spoken throughout the north of India by so many millions of people that it seemed churlish not to give it a go. So I bought a book and sat down to learn. After a few hours I ditched the notion of learning the new alphabet, I had become so bored with it that the words were swimming on the page in front of me. I instead elected to try to learn phonetically, why not? Hindi is written 'in English' all over India, and while no two spellings are the same; there must be some pattern?

I started by trying to decipher some Hindi film songs that were my particular favourites. I listened to Asha Bhosle's all time classic "*Dum Maro Dum*" and transcribed what I thought I heard. The first word stumped me; *dum* apparently means both 'puff', 'pressure' (as in cooked under pressure or quickly as in a pressure cooker) and 'power'. *Maro* means 'take'. I guess the title means 'puff, take a puff' since the film it is from is about hippies and marijuana, but the alternatives were too ambiguous. I tried another classic, Mohammed Rafi's barnstorming "*Aaj Mausam Bada Beimann*". I already knew *bada* meant 'big', or at least I thought it did, but after some painful detective worked and some applied logic, I surmised the whole to mean 'Today's weather is a cheat'. I immediately gave up on the phonetic translations.

Determined, I found myself a phrasebook entitled 'Quick and Easy Way to Learn Hindi' which sounded right up my street. Given my tendency towards embarrassment with languages I am very wary of phrasebooks, particularly when remembering Monty Python's Hungarian one, but I bought it nonetheless. What I found inside was utterly useless but truly magnificent. The conversations, which are presumably supposed to follow relevant patterns, have a futility and peculiarity which beggar belief.

At The Hotel

"Darling, have you ever seen a cabaret?"
"No, dear."
"Get ready, I'll show you how."
"The cabaret is nice or not?"
"It is very nice."
"We'll take dinner there too?"
"Sure, Tandoori Chicken."
"You'll drink too?"
"A little, a peg or two."

Having Bed Tea (a cup of tea in bed before breakfast, one of Indian life's little luxuries, often an option in hotels).

"Good morning dear."
"Good morning sweety."
"Have bed tea."
"Thank you."
"It's nice?"
"Yes and you are nicer."
"It's hot and you are hotter."
"It's having flavours isn't it?"
"It is and you are having more fragrance."
"One should learn from you how to praise."
"Beauty deserves praise."

Worried that such conversational ability would get me into trouble with taxi drivers, shopkeepers, etc, I moved on to more direct word learning. At least I would be able to say 'turn left', 'stop', 'hurry up', 'how much is this?'

Nothing of the sort. No mention of any useful words at all, not even 'hello'. Plenty of phrases to demonstrate prepositions and conjunctions such as "He will die within a week" (*'vah ek saptaah men mar jaaegaa'*) and the essential "No sooner did he get up to deliver a speech than the hall resounded with clappings" (*'jyonhee vah bhaasan dene ke liye uthaa tyonhee bhavan taaliyon se goonj uthaa'*), and plenty of archaic technical terms and complicated adjectives. So I could say 'nomenclature' (*shabdaavalee, naamaavalee*), 'agrarian reform' (*krishi sudhaar*), 'pendown strike' (*lekhneeband hartal*), 'sex perversity' (*apraakritik kaam vaasnaa*) and of course 'refugee liaison officer' (*sharnaarthee sampark adhikaaree*) but had no idea how to say 'two cups of tea please'. Clearly this wasn't going to work either, in fact I learned 'he will die within a week' and tried to say it to a few work colleagues and they could not decipher a single word, reverting instead to raucous laughter.

As a final attempt I began to ask work colleagues to help out by asking either 'how do you say…?" or 'what does…mean?' whenever relevant. On ninety percent of occasions they either didn't know (I once called Sarasvat from a roadside mechanic's close to where my car had broken down to ask what the word for 'spanner' was, he didn't have a clue, 'maybe it is *'pana"* he said, 'the best thing is to just motion as if you are undoing a bolt'), or I was told 'you can't really say that in Hindi' or 'it doesn't really mean anything'. So I gave up. What use is a language that cannot be used to express what one means and which 'doesn't really mean anything'? So I had to resort to learning Indian English instead.

Indian English is more or less the same as English and there is no need to make any effort to learn any specific words. The difference is mainly in the tone. Most Indians speak English at a very fast pace and with a very 'Indian' accent. In an effort to be understood I adopted the same approach, and to my surprise and total embarrassment, it worked. Adopting an entirely condescending, Jim Davidson style Indian accent, wobbling the head while talking and clasping the hands in a *'Namaste'* position I found that people suddenly understood me. The revelation came to me via a rickshaw driver after the following dialogue:

"Hello, North Main Road please."

"Huh?"

"North - Main - Road please." Slower and more deliberate, really sounding out the vowels.

"Huh? Kya? Kya?"

"Noorth - Maain - Roaad." Shouting as if the man was a complete idiot.

He looked at me in utter confusion and it suddenly came to me, try an Indian accent.

"Nort Mayn Rood." Clipped short and spoken as fast as possible, trying to sound Indian and wobbling the head in a "Thousand apologies" 'Mind Your Language' way.

"Ah, Nort Mayn Rood," his face lit up in recognition. "Nort Mayn Rood, no problem, why you not say before?"

And so it was that I began to go about my business talking to people as if I was taking the piss out of them.

"Hallo sir, I am wanting to be buying some of your excellent Indian tea, which shelf of the shop will I be finding this?"

At work this became very popular and I was put on the spot in front of all sorts of people with 'have you heard Mick's Indian impressions?' after which I would be forced to say something like 'I am finding too much of cold in this office, A.C. is turned up so high, not like in U.K.' and they would laugh at this.

I had to bring it all to an abrupt end when one of my staff offered my impressions to some English clients. I of course refused saying that it would not be funny and tried to get out of it, but the pressure mounted with more and more staff saying 'oh, its so funny', 'come on Mick please do some impressions' I winced more and more. 'At least do the 'Oh Bloody Hell'" said Shanti wobbling her head and accentuating her own accent. The clients looked at me in horror and the scene became more excruciating as I tried to explain to the English people that this was appropriate between friends and had started through me not being able to communicate without sounding like Jim Davidson doing an impression of an Indian. I managed to climb down with my reputation intact to a degree but was very careful not to use the accent as a joke in future. Being English, the clients knew that this was ultimately racist humour in the worst possible taste, interesting that the Indians themselves did not see it this way at all.

****

As May turned into June, the heat became more and more intense, and the promised air conditioning unit had still not materialised in the flat. Finally, after semi-violent remonstrations, it arrived.

The Indian love affair with the acronym is well documented. Read any newspaper and you'll be obliged to make a note of the first few letters of every institution or official body mentioned even when the establishment in question is referred to only once throughout the entire piece. Randomly from today's Times of India - the London Contemporary Dance School (LCDS), the Symbiosis Institute of Management Studies (SIMS) and Inter Services Public Relations (ISPR) whose Initial Publicity Officer (IPO) was being called to question. In all of these cases the acronym was not repeated. Acronyms are everywhere and can be a real pest.

When I first visited India I checked into a hotel and was propositioned thus:

"A.C. or Non A.C.?"

"I'm sorry?"

"A.C. or Non A.C. Sir?"

"I'm sorry, I don't understand."

"A.C. sir, you want A.C. no?"

I shrugged my shoulders, confused.

"In the room sir, you want A.C. no?"

"I have no idea what you are asking me so I will say no."

"Is ok, fan is there."

Now I was totally confused. Upon entering the room I saw the unit and the penny dropped, A.C. meant air conditioning, which, if opted for, ads extra to the room tariff.

This is fair enough, we all use PC for personal computer and F.C. for football club, so in a nation of intense heat, A.C. is a very apt term for air conditioning unit. The problem is that A.C. refers instead to Arctic

Conditions. Air conditioning in India is either broken, not working due to a power cut, simply off (and thus making the room smell a little off) or VERY MUCH ON. Going into a bank cash machine booth or ATM is like entering the cold store at a frozen food store, air conditioned pubs are so cold that it is advisable to sit near the toilet because you'll need to go three or four times for each beer you drink, air conditioned restaurants should never be attempted without a coat, hat, scarf and gloves and the office in which I was stationed was giving me chilblains. A.C. in India is fine as long as one is at leisure to regulate it oneself, leave it up to anyone else and you're ice cream.

But summer prevails and A.C. at home becomes essential for sleep so as I heard Theme From Love Story playing and the doorbell was pressed, signalling the arrival of the A.C. unit I breathed a cool sigh of relief. I opened the door to reveal three grinning faces, a filthy broken piece of nineteenth century machinery and a fourth man grappling with the lift door, trying to shut off the deafening Theme From Love Story. When he finally succeeded, he spoke in the sweet silence, "A.C. sir A.C?" He pointed to the twisted metal on the floor. I ushered them inside nonchalantly, aware that another fiasco was beginning. The four tiny men picked up the contraption together, fitting perfectly around its one metre width. Their filthy hands gripped the device and shuffled it through the flat, and an eight legged air conditioning unit installed itself in the hole in the window frame. It was switched on. Nothing happened. The men left anyway, seemingly satisfied that their work was done. On the way out they were careful to touch the newly painted walls with their dirty hands wherever possible. I closed the door after them and called the landlord.

"The A.C. has been installed but it doesn't work."

"OK, I'll send another tomorrow."

Another sleepless night.

The following day the sequence went something like this: Theme From Love Story, 'ding dong'. I open the door to five dirty men, one trying to make Theme From Love Story stop. He succeeded and they entered as a unit, four in a group, one bringing up the rear. They removed the broken unit from the frame, became the eight legged A.C. creature again and exited, the extraneous fifth man touching the walls with his dirty hands on behalf of all the others. Theme From Love Story plays and stops.

Five hours later, Theme From Love Story, 'ding dong', 6 men, 5 small, 1 large. They have a newly broken A.C. unit with them, which they shuffle in, install, dirty the walls and try to leave. I stop the new large one and get him to turn the machine on, he defers to one of the original small ones who does so. It blows air but it isn't cold. They leave. Theme From Love Story. I call the landlord.

"Another A.C. came, it is no good, it doesn't work. Please send a NEW one, this one has already given over thirty years of service."

"OK, sorry, I'll send new one tomorrow."

Sleepless night.

Theme From Love Story. 'ding dong'. Seven men (yes seven), a *less old* A.C. unit. Oh, fill the rest in yourself, but this time it works. The seven dwarfs (ok six and one comparative giant) look happy, and as the room gets colder and colder they set about touching the walls. I too am happy and I tell them so. Really I should hate them for this ignominy but I don't, this is how things happen here, maybe I'm finally getting used to it, maybe my irrational bouts of anger will stop? I thank them and show them out. Theme From Love Story.

With the A.C. working in the bedroom, sleep became far easier, both the temperature and the hideous noise it made were helpful. The noise helped to drown out Theme From Love Story and the intolerable sound of reversing cars. As Indians drive with their ears, and pedestrians stay alive by using their ears rather than their eyes, there becomes a need to have reversing noises attached to all vehicles. The 'stand clear, this vehicle is reversing' recording can be heard on occasion from buses and lorries, but private vehicles have merry jingles played at deafening volume with which to warn pedestrians of their imminent encroachment. At times India resonates in the way I imagine a convention of ice cream men would, and the tunes are always grating, happy tunes, like comedy ringtones but without a trace of irony. Our neighbours all had such tunes and as they came and went at all hours of the day and night we were treated to 'Happy birthday to you', 'We wish you a merry

Christmas', 'There's always Coca-cola', 'Barbie Girl' and amazingly, 'Theme From Love Story' meaning that on occasion, when serendipity permitted, the car and the lift would coincide, and we'd be treated to a warped stereo version of this recurring refrain. Just below the bedroom window was a small floral roundabout which one neighbour felt the need to circumnavigate in reverse on each occasion of his arrival or departure. As he circled, an incapable speaker would belt out 'Silent Night', usually getting at least as far as the second 'sleep in heavenly peace' before stopping.

Two days after the A.C. was installed and working, a power cut was well under way (on Thursdays, Pune is subjected to all day power cuts which sometimes continue through to Friday afternoon. This regular power cut off is apparently necessary to provide time for the electricity to be 'fixed', as I have said before, the electricity situation is beyond my comprehension) and the night was hot. At 1am the neighbour returned home and I was treated to the unambiguous opening bars of 'Silent Night'. Within one second I had the window open and was shouting "It would be a fucking silent night if you'd stop fucking reversing, just drive forwards into your parking space like everyone else you dumb inconsiderate twat." The irrational bouts of anger were back.

Except when the electricity would go off, the flat became a place of refuge from the sun for both us and our menagerie of wild pets. Julie's persistent feeding of the mynahs resulted in a pair nesting on the window ledge. We quickly named them Morris and Ursula and looked forward to seeing their chicks (they didn't have any in the end, we couldn't work out why). Julie was able to entice Morris in through the window after only three days and as I returned home from work each day I readily expected to find the three of them sitting on the sofa, chatting and having a cup of tea. In addition to the mynahs, red-vented bulbuls started to visit and on numerous occasions, they too came into the flat and would perch on the outstretched extension lead (there's never a plug where you want one) and sing away. The palm squirrels would perform death defying stunts, chasing each other across the sheer face of the buildings at impossible speeds, while the house swifts and dusky crag martins would swirl about them, the martins occasionally resting on the eastern window ledge (a rare treat, martins almost never stop flying and have to lie on their bellies to rest). But my favourite guest was a tiny, irritable and erratic baby gecko that would hang around the lift, looking for flies and listening to Theme From Love Story.

Julie also began art classes that were largely taught in Hindi. The only Hindi she knew was *'teekay'* (OK) and *'Kulja Sim Sim'* which means 'Open Sesame'. Each day I was expecting her to return with lots paintings of Ali Baba and the 40 thieves. She also engaged further in making contacts in the sparse Pune art world, people who, although thin on the ground, seemed enthusiastic about getting her involved in whatever came up. So things were settled and happy again, and despite the occasional outburst, we were fairly happy and enjoyed exploring the new city. We spent time socialising with my workmates, trying all the new pubs and restaurants springing up all over the city, getting to know the student areas and watching so many young people studying each other while pretending to study. In fact we spent a good two months investigating the city before the realisation hit us that, actually, considering how exciting it should have been, we were quite bored.

First my work became a little boring, after two months I hadn't really done much. The novelty of working in India had worn off and the day-to-day repetition of the working life had begun to surface. The workload was not increasing at the pace we had anticipated, some clients had got cold feet over the whole outsourcing issue and the impetus was lost. Excuses were being made about why things were taking too long, but we all knew that the real reasons lay in the backlash to outsourcing that companies were having to deal with, particularly in the US. It became clear that while the migration of such work to India was inevitable, the pace at which this happened would be greatly affected by public acceptance, and for now, the public did not accept things at all. The slow progress meant that we had time to spare in the office and were not under pressure to 'ramp up' the processes quickly, so could spend some fairly lazy days discussing the future. While the notion of lazy days at work may seem nice, the reality is a sense of futility and waste that I find very hard to deal with. When there is no work to do I would rather not come to work, but this was deemed unacceptable, despite my protestations, and many days were spent by the whole team (around thirty people) *pretending* to work. This clock-watching hell was leading me further and further into angry ennui, and Julie's efforts to find some artists to work with had also proved fairly fruitless, so a new type of frustration set in. We discussed backing out and heading back to Goa, even just heading back to England and giving the whole thing up, but sense prevailed and we determined instead to stop moping about and use the free time constructively. We would stop trying to find third party leaders and become our own masters again. We would write and paint at home, get out and about more, buy a car, buy a guitar, take up some classes and above all cheer up. As an affirmative unveiling of our new happier selves we would throw a party.

I had no idea how to have a party in India and needed to find out the protocol by asking my friends at work. I was informed that it would be expected that the host would provide food and drink, that it should start quite late, say around 10pm and that it would go on until around 2am. In terms of guests, I felt that the flat and roof terrace would comfortably hold fifty and on the assumption that only fifty percent of those invited would come (again a figure quoted to me from an experienced source), I decided to throw the invite open to absolutely anyone and everyone in the office. I picked a Sunday night; an unusual night for a party but Sunday night is the new Friday night for many people whose work entails talking to people over the weekend as they have Monday and Tuesday off and return to work on Wednesday. I was slightly concerned that a rowdy party on a Sunday night might cause upset within the housing colony, and asked Sarasvat if he felt it would be ok. He tried to reassure me by telling me that he had had plenty of such parties over the years and whenever complaints were lodged, he simply reminded people that the property was his and that he could do what he wanted in his own property. I was doubtful that such a supremely selfish defence would appease anyone at all, but gave in on the basis that there really was no other day possible for the majority of the people I worked with, and I also felt like giving the neighbours back something for their weekly *pujas* that rattled the ornaments on my shelf.

I invited everyone I knew at the office, first circulating emails and then in person. When Shanti got wind of this she chastised me demonstratively.

"Noooo Mick. Don't just invite anyone, you will have some really horrible people turning up, people with nothing to say and no idea how to party, yaar? We must be selective about who comes, leave the inviting up to me, yaar?"

"No, I'll invite who I want, the more the merrier, I don't just want a circle of friends, I want to get to know people and I want people to get to know me."

"Noooo. People will get to know what you get up to, they will talk about it in the office, you'll get a bad reputation."

"It is of no consequence to me what people think about me in relation to how I behave socially, for God's sake Shanti, surely you know I don't care about that. Let them talk. I will relish the attention. I thought you knew me better than that?"

"Yaar, OK, fair point, but if you're not careful people will turn up with their families, kids, mothers, even brother cousin uncles, y'know? You don't want that?"

"You think young people will bring their mothers to a party? Surely not?"

"For sure, some will not be allowed to go if they don't. And their mothers will not approve of drinking at all, it would be terrible. No?"

I quickly pictured a scene where drunk teenagers were being chastised by irate women barking wrathfully in Punjabi, furiously wagging long index fingers in my direction.

"Err, I never considered that. Well, I'll bow to your better judgement, but don't exclude anyone on the basis that they might not normally 'party hard yaar' and are not into 'really freaking out' or 'having a blast' OK? None of these terms that you all use so frequently come even close to describing the real situations they try to describe anyway, I want to have a good laugh and am happy for anyone to come, anyone OK? Just not their mothers."

And so it was left up to Shanti to bring the people. She reckoned about fifty would come and I was very happy with this. I asked how much drink I should provide, suggesting a couple of hundred tins of Kingfisher and a few bottles of spirits and mixers. Apparently this would be far too much, as many people would only have one drink.

"One drink all night? I'm not having that. That is nonsense." I was appalled.

"Of course some will drink more, Indians are not too good at beer

Mick, come on, you know that. Get more bottles of Bacardi, then people can just drink Coke and pretend to be drinking spirits if they feel too drunk. Most of these people are kids Mick, they are under twenty, you'll have teenage puke everywhere if we're not careful." Manohar said, putting me right on a few issues.

"I very much like the way you said 'you'll' have them on your hands, I'm making you responsible for cleaning puke as I'm taking your advice on alcohol."

We installed Shanti's midi system as my own traveling music set up had inadequate power. The speakers overloaded a little and everyone seemed to be happy with that. Julie and I spent two days buying and preparing food. We had noticed that canapés were all the rage and went into canapé overdrive putting at least one of everything available in the supermarket on to hundreds of little biscuits. We also made a few English classic staples like roast potatoes, strong pickled onions (I'm a keen pickler) and chicken liver pate for the non-vegers, as well as buying in a few safe Indian bets like *wadas* and samosas. With all this on the table we turned up the music and waited for the doorbell to ring.

The first guests arrived and I proffered beers. Only one was taken, the others saying 'I'll have a drink later'. It was a similar occurrence from the next ten guests who all arrived together. Everyone came bearing gifts of sweets or flowers, and hardly anyone looked like they were going to drink. By eleven o'clock, there were forty-two people in the flat and only twenty of the hundred or so beers I had rammed in the fridge had been drunk, the spirits bottles too were largely still sealed. Despite this, people were definitely having a good time. There was dancing, (well at least there was now that I had relinquished the music system to a younger, better looking, *desi* DJ, Shoabh, who had had the foresight to bring a case full of *filmi* hits - there had been absolutely no interest in the Ramones, and the Fall just brought vehement booing) there was lively chatter and the beginnings of horseplay on the balcony, and the formalities and small talk had clearly been dispensed with. Around twelve o'clock I checked the fridge again and WHOOAHH there were only about thirty tins of beer left. Fifty tins in one hour? How did that happen? I looked around the flat and yes, everyone was drinking and there were plenty of empties, I realised fifty beers an hour for forty odd people is hardly liver bursting consumption but felt happy that things had picked up a little.

I got a whiff of grass and located its source. I am not much of a dope smoker, but I have my moments, and now was certainly one of them. As I approached the joint manufacturers, I sensed a caginess I had not expected from two men I had first met some eight months before.

"Let me in on that then." I interjected.

"On what?" Asked Nilesh, concealing the joint behind his hand as if he were behind the bike sheds.

"Oh please, give me some credit, I am thirty five years old."

He produced the joint and asked me, "Do you dope?"

I laughed at the awkwardness of the phrase. "Doesn't everyone?"

"No, definitely not, don't tell Sarasvat or any of the other bosses about this." He looked genuinely worried as he passed the joint to me. I called Sarasvat over, "Bit of *bhang* going over here if you want some."

He came over and took the joint immediately. "Good, good, any more where this came from?"

Nilesh looked shifty and Ajit tried to walk off.

I had to straighten things out, I had teased them enough. "Gentlemen, this is fine OK? Smoke as much as you like, tell everyone I gave it to you. No one is going to mention this to anyone at work OK? Just relax."

The music crept up louder and louder and the dancing became more vigorous, at least among the women. I noticed a small group of men flicking through CDs in a fairly enthusiastic manner and suddenly the music was cut and the dancing stopped. Groans came up from the dancers as the aforementioned men took the dance floor and we all realised the inevitable was happening. As the first awful notes rang out and the air guitar stances were assumed, I recoiled to the balcony with my hands over my ears, but the irrepressible sound got through as it always does.

"OOOh, I got my first real six string…"

I saw Julie cowering on the other side of the terrace with her head in her hands, she looked up at me and shouted across, "Is this really happening?"

I nodded, it was really happening, 'Summer of '69' by Bryan Adams was being played in our home. There were no two ways about it, the song was actually being played in our own home, our own sanctuary and refuge from the horrors of contemporary Indian life. The impossible had occurred, and was still occurring right in front of our panic stricken ears and there was nothing we could do to stop it.

There are many things that travel teaches us and many unexpected situations are thrown up in the name of adventure, but had you asked me if Bryan Adams would ever be played in my home I would have bet my life against it. But here it was, and in fact, it wasn't really *that* bad. Oh, it was horrible, make no mistake, but it didn't kill me and apart from overwhelming anxiety, I was actually unscathed, at least for now (there is of course no knowing what the long term psychological effects might be). The volume went up and up as more rock horrors ensued, and just as I was beginning to think of throwing the switch and faking a power cut, we were saved by the bell.

"Ding, dong, ding, dong, ding, dong, ding, dong, ding, dong."

The music stopped dead and looks of fear were traded about the room. It was nearly 2am and this was clearly not another guest.

Someone opening the door and a small, shaking old man hopped into the room on anxious stiff legs like a hostile robin. He tried to shout but his wide-open mouth made no sound; he motioned his anger with flapping arms. I stepped forward with authority and ushered his rigid little body back over the threshold, where outside the door in the stark light of the stairwell I saw the security *goondahs* for the first time.

The little man's eyes rolled as he began to spit his tirade of bile at me and I realised he was drunk. I felt pity for him, he had clearly spent the evening becoming more and more enraged at the noise, drinking more and more and working himself into more and more of a frenzy. As he vented his frustrations I smiled at the seven or eight security men with their *lathis* in hand whom he had obviously rounded up before finally having the Dutch Courage to intervene. I nodded at the two men who I said hello to each morning on the gate and they betrayed their leader by nodding firmly back and smiling at me as they always did. The scene reminded me of Dad's Army, only with an emaciated and less restrained Captain Mainwaring with his full subdivision of unkempt misfit soldiers goofing around in the background. As he spat through his invective I picked out quite a few choice swear words among his drunken ramblings about how disgusted he was, how the colony rules of 10pm curfew were being grossly transgressed, how depraved and decadent us foreigners were, how he had lived in Australia for five years but hated the lack of discipline among *gauris* (whites) and how I was leading the Indian youth astray with my debauched merrymaking. While silently absorbing his vitriol, Sarasvat and I had been politely patting the security guards on the back saying, "it's ok, no problem, nothing to see here, move along now, the party is over", and trying to usher them downstairs. In the main they seemed quite happy to move off, having become a little bored with the lack of physical action and unable to follow the conversation which was being conducted entirely in English. Finally I addressed the pugnacious pensioner thus:

"Sir, I would like to wholeheartedly apologise for the upset my friends and I have undoubtedly caused you. We will stop the music immediately and cause you no further distress. In future however, I would suggest that you knock on the door before you have worked yourself into an unhealthy state and simply ask politely for the music to be turned down. You will find that we are quite approachable, and quite the opposite to the immoral savages you have taken us for. In addition, I would ask that you do not bring your band of apish *goondahs* to my door again, unless you actually want a fight, a fight which, if you look around you, you and your band of dishevelled old men are very much set to lose, *lathis* or no *lathis*."

Anywhere else in the world I would have been punched in the face half way through this delivery, but here, it really worked. He simply nodded, turned and walked back down the steps, followed by his band of grinning goons, each of whom Sarasvat and I took great pains to shake hands with as they passed.

We went back inside and turned the music on low. Two minutes later there was a mysterious power cut which occurred only in our block; music over. In truth I was relieved, while I'd stood my ground with the complainant, there was no question that we were in the wrong, keeping everyone within a kilometre radius up on a Sunday night. Then of course the singing started. Put more than five Indians under thirty years old in a room and within seconds they will be in song. After a couple of pitiful tuneless ditties it all looked like blowing over when Ajit returned from the toilet clutching my new acoustic guitar.

We all knew what was coming next…

"OOOh, I got my first real six string…"

AAAAghh.

Chapter Thirteen

In the office there was less and less work to do. The team had been trained on everything possible and we found ourselves sitting around doing far too little to satisfy anyone's expectations. The party had been a nice diversion, and the knock on effects had been quite hilarious - I had apparently got myself quite a reputation as rumours flew about what had happened. The old neighbour had complained to the colony manager who in turn complained to my landlord, who in turn complained to my company admin department (they had finally brokered the deal and the rent was subtracted from my salary) who in turn complained to me by email. I was furious. I flew into the office and tore a strip of the head of admin (what a pleasure) telling him it was my business what I did in my own home, that I am a fully grown man, that he was not my father and that if he did not apologise to me formally in writing, I would resign and he could answer to my boss for that. The next day I received a letter of apology, which I would love to print here, but I think I've done enough juvenile gloating as it is.

I started not bothering coming into the office and told my colleagues to do the same. I was making a point. I called Avinash and told him he would find me at home if he wanted me, and that until there was something to do I would not go to the office and neither would the rest of the team. I told him I was bored stiff and reminded him that I was not doing this for the money and that if there was really nothing to do I would sooner just resign and go back to my original plan to ' tinker about with old cars, play guitar, get drunk, sunbathe, etc' back in Goa. Frustratingly he agreed with me, there was nothing he could do about the lack of work situation either. Thankfully, less than one week after my proclamations to Avinash, another client expressed interest and we got the go ahead to train new staff and redeploy or further utilise the ones we had already trained. This would be within the next month. In the meantime, I was able to spend more time with Julie and get out and about a bit as planned. The first thing to do would be to buy that car I had promised.

As I have mentioned before, finding vintage cars is a tough job in

India. They are relatively numerous, the problem is in explaining why you want an older car in preference to a new one. Each time I asked one of my colleagues where I might buy a Premier Padmini, the immediate response was, 'no, you don't want one of those, they are no good, you want a Maruti 800'. Tired of justifying myself I found out where the second hand dealers were in the city and set off to find a suitably excellent motor. I embarked upon chasing three or four wild geese, which led only to the inevitable Marutis, before stumbling upon East Street in the Camp area of town (It is called 'Camp', but as far as I know it is not populated by drag queens or anything). This ancient avenue which I came across one night while trying to locate the indispensable 1000 Oaks English Theme Pub with a rickshaw driver who was stopping every third person and shouting 'Towsand Ox?', runs parallel with MG Road and boasts Pune's oldest cinema in addition to hosting the main dealer for the imperative Hindustan Motors (there is also a *paan* stall which is touchingly titled Wife of Disabled Dewan Kiosk). It is also home to a number of shabby car lots full of atrophying vehicles from former ages, and it was in one of these that I spied a 1962 Triumph Herald in tip top condition among a family of Padminis, Ambassadors, Chryslers and if I was not mistaken, a 1950's Bedford van (the one that looks like a huge piggy bank). The very next day Julie and I paid this holy site a visit and introduced ourselves to the perspiring proprietor, Mr. Khan. We shook hands heartily and I expressed an interest in the Herald. It was too expensive and I wished I wasn't white. I offered some half-hearted bartering but he was resolute.

"This car is not just for anyone, the right person will pay the right price. None of my cars are for sale to people who just want a cheap car, see, these cars are special cars."

What a salesman, this was a language that I could speak. We looked around his lot and I asked Julie whether she liked the Herald (or Standard as it was known in India). She didn't.

"It looks like the Ghostbusters vehicle. It is hard enough being white and female, imagine the looks I'm going to get if I have to drive that about."

Fair enough, the shape is similar and it did have a tapered stripe down both flanks and a doctor's sticker on the windscreen, not to mention the two large eagle stickers on the wings where the Ghostbusters logo would be. We turned our attentions to the Padminis. They were mostly in a sorry state, but there were a few possibilities. After so long without a car, I wasn't willing to wait any longer or pass up this opportunity and I spoke to Mr. Khan again to that effect. I explained our predicament and that we were looking for a Padmini but wanted something in better condition to those he had available.

"I don't think you understand Mike…"

'Mick,' I interrupted, "My name is Mick."

"Sorry, Mike, I was saying that these cars are available at a very cheap price, but for an additional sum I will completely restore them to their original condition. This is what I do."

"How much are we talking about?"

"Well it depends which car you start with, the restoration, new paint, chrome, fixtures, interior, engine recondition, everything should cost less than ten thousand rupees. Come, I will tell you how much for each car."

Shortly the deal was done. We chose a later 1989 model that still wears the same 1970's body, applying the logic that it would look old but drive like new. It had been previously owned by an Army Major, which is apparently a good thing, and the total cost for buying and restoring the vehicle would come to around twenty five thousand rupees. I expressed concern about the reliability of such a car and Mr. Khan assured me that if it were ever to break down in the city, I should just call him and he would be there within half an hour to fix it. He also assured me that if we did not like the vehicle or no longer wanted it at any stage, he would purchase it back at a depreciation of one thousand rupees per month. This was a deal I could not pass up; we had paid five times that amount per month to rent the car in Goa. It would take around three weeks to complete. Knowing how long three Indian weeks can take, I persuaded him to loan me the Herald at the same price of one thousand rupees per month until the new car was ready. He agreed. Mr. Khan is a unique businessman, exemplary in every respect.

As we drove off in the short-term comedy vehicle, Julie was visibly less than overjoyed with its slapstick appearance. She vowed not to drive it herself and fairly quickly vowed not to drive in it with me either. I was of course thoroughly happy with every aspect of it. It was a particularly ridiculous vehicle - it was too small, too slow, too hot and mostly it didn't work. But it had a horn so loud that it made other people fall off bicycles, an air assisted klaxon of the OOEEERRGHH, OOEEERRGHH school. Driving it was like driving a cake. It had over sprung suspension and ineffective steering and was simply horrible to be in. It broke down three times in the first two weeks, on two occasions it cut out on the main road (four lane traffic) as I was turning right into oncoming traffic. Indian traffic does not stop when something in front stops; it simply goes round it while sounding the horn. This can be most disconcerting when you are the 'it' in question. Altogether this was a hopeless car, but Mr. Khan was true to his word, turning up on his scooter within twenty minutes each time it stopped working, fixing it and driving off without complaint. He also never apologised for the car breaking down which I loved; I had become very tired of the subservient nature of some Indians, and was pleased to meet someone with a clear idea of equality in life.

"Mike, this is a very old vehicle. Old vehicles break down. If you want a car that does not break down, buy a new one. Oh, and don't buy an Indian one."

I could now drive to work, and thoroughly enjoyed the commotion I could cause by sounding the horn outside the car park gate and watching the security guards jump out of their skins. I was aware of the hypocrisy of me using a horn, but what the hell? This was *funny*. My workmates were partly astonished at my vehicle but mainly embarrassed. I could not persuade these yuppies to even sit in the car, but I did not care a jot. *This* was my idea of fun.

****

Knowing that it would be a while before we had a vehicle capable of doing any mileage, we decided to take a bus to one of the surrounding hill stations for the weekend. After some deliberation we settled on Matheran. Hill stations are another legacy of the British Raj. Needing refuge from the unforgiving climate and disease epidemics which ravaged the ranks of the early settlers, the Britishers built numerous outposts anywhere they could find more temperate altitudes. In their heyday, these settlements provided seasonal retreats from the searing cities in summer and the holiday atmosphere of so many toffs converging on so few places gave rise to a heady comportment of imperial pomp and aplomb. The hill stations comprised grand colonial houses with lawns and gardens where one would spend the day playing croquet or taking tea and cucumber sandwiches; a miniature England of green rolling hills; the Highlands in India's high lands.

Today, the hill stations still display a peculiar Britishness which is evident not just in the climate, the panorama and the faded glory of the grand old houses, but in the fact that these holiday retreats have kept up with the trends of the great British holiday and somehow metamorphosed into mini Blackpools. Nearly all of the hill stations have a central pedestrianised strip flanked by candy floss sellers, *chikki* sellers (a brittle honey-toffee - the Indian version of seaside rock?), donkey rides and penny arcades; the promenade reconstructed on the top of a mountain. And like in Blackpool, in summer time, one can expect a stroll along said promenade to be an unpleasant affair, noisy, congested and litter strewn, as it will be. But get out of the crush and into the wilds and these places outstrip anything the British seaside can offer. Great vistas of unpolluted air, breathtaking views, mild mountain breezes, and an uplifting sunny day every day.

Of the six or seven hill stations still popular in Maharashtra, Matheran (meaning jungle top) is perhaps the most incredible. A tiny toy-town, built some 800 metres up, it is invisible as the mountain is approached from any angle. Indeed, the mountain is almost sheer, and the climb by bus is pants-droppingly terrifying. The little village perched on top of this awkward monolith is accessible only by foot, by horse or by narrow gauge toy train. Motorised vehicles of any description are barred, putting Matheran in the unique position of being easy on the ears.

The bus left us some four kilometres from our destination and as we alighted with our weekend size packs we were set upon by a swarm of climacteric *coolies*, hand cart pullers and pony wallahs. Fortunately for us, the toy train was pulling in and we were able to eschew the time-honored methods of muscle power in favour of the progressive propulsion of steam. As the little engine climbed, the langurs swung menacingly in and out of the bogies like, well, like monkeys really, baring their teeth and laughing as we recoiled in fear, pretending not to be bothered.

We checked into a marvellous little hotel whose grandiose Parthenon façade concealed four petite en-suite rooms behind which lay nothing at all, a metaphoric representation of this veneer of a town. The next morning we checked into Lord's Hotel where we had been able to get a reservation for only one night. Lord's is a flawless example of paling Raj glory. The mountainside pool, the crumbling chalet accommodation and the regal dining room of the main building spark an atmosphere of worn grandeur that is hard to beat. Full board is the only option, including bed tea, breakfast, lunch, dinner, and supper. Lunch is four courses while dinner is five. The gold thread embroidered tunics of the elderly waiters catch the sunlight streaming in from the west through the huge viewing window and shine in a way that somehow says 'the world is beautiful, everything is alright and there really is a plan'.

It was hard to leave the hotel, but that evening we braved the red dust of the prom' in search of a bar. While searching we were buttonholed by a pretty little girl who offered "Small slippers sir?" We bent down to view a little cardboard square onto which had been stuck ten pairs of tiny slippers fashioned from vibrant paper, glitter, thread and presumably a good deal of effort and patience. I pulled out a wad of notes, very happy to but something for once.

"They are beautiful, I'll take them, how much?"

"Five rupees please sir."

Why is it always that the best, most intricate and beautiful, and most unique souvenirs are always offered at such paltry prices by the nicest, most honest, poorest people in the world? I gave her twenty rupees. Well, you take a bargain when you find one don't you?

We found Promod Bar and ordered some beers. After dinner at Lords we were stuffed to the gills, but this was a holiday and consuming would be at a premium. The tiny bar was the usual men-only affair and Julie's presence created typical interest. We got chatting to three men from a diamond mining company somewhere in Gujarat (a dry state) who were celebrating a big contract they had just won somewhere in the Gulf. They were at pains to swagger their wealth and produced huge diamond encrusted rings for our approval. I offered my small slippers as a rival in beauty and suggested a straight swap. The laughter that rang out as they realised what I meant was far too loud for the bar and the landlord came over to check everything was all right. A quick chat in stifled English along the lines of 'its ok we're from Gujarat' seemed to clarify the situation and the landlord returned to his bar. I guess Gujaratis must be culturally allowed to make more noise than anyone else.

Later on the call centre crowd arrived. Matheran is probably the closest hill station to Mumbai, so attracts plenty of young office workers up for a short break to buy hats, drink beer, watch the sunset, shout a lot, and then return back to the city happy. We got chatting to three chaps with a familiar story. Good education and family background, two years taking calls, promotion and the inevitable strive for 'growth' within the company. I told them to quit, get out of the trap before it was too late, help out at a tiger sanctuary or something, do anything but don't slave away for foreign companies at vastly reduced rates. They laughed it off in their well-dressed, confidence-in-youth manner and I let it go too. In fact I wondered where the outburst had come from. Was I changing my tune? Was I turning against offshore outsourcing so soon after justifying it so soberly? I think it was the realisation that office work is more or less identical regardless of where one carries it out, these young men could have been myself ten years ago, downing pints in the City of London with the Bulls and the Bears, and somehow this made me a little sad.

Our luck was in and we were able to stay another night at Lord's. In celebration I agreed to accompany Julie on a horse ride to Sunset Point to witness the well-documented spectacularity of the sunset. I am both a coward who is troubled by horses and a coward who is afraid of heights. This trip would amalgamate both fears, and as twilight approached I insisted we nip into Promod Bar for some Dutch Courage. Two large Kingfishers and I was ready. Julie is a proficient horse rider, indeed, she has competed in dressage and show jumping to a distinguished level. I allowed her to select a pony for me and a horse for herself (they're probably all ponies but they looked like Shires to me – gigantic, uncontrollable, ill tempered beasts). She jumped on hers and I gingerly and clumsily mounted mine with help from both its owner and two passers by. I am a man and I am not afraid to admit when I am terrified.

"Julie, I'm terrified."

But she had sped off into the distance, leaving me frozen with fear and at the whim of the moody beast I was astride. I pleaded with the horseman, "Don't let go" as he let go and whipped the horse on its behind, sending it cavorting towards its partner. "WHHOAAH," I tried, remembering the westerns, "Stop, Horse, Stop, Stop." I screeched but still it sped on. As the animal reached Julie's horse I was clinging on with terrified hands. She stopped, expertly manoeuvring her horse to an elegant standstill, "Pull on the reigns hard" she instructed and a reflex I never knew I had did this for me instantaneously. As my hands pulled, the horse stopped dead and reared up onto its hind legs in a 'Hi Ho Silver' style, throwing me into an involuntary horizontal, slipping-off-the-back position. As it reared I glimpsed the sheer drop at the path's edge for the first time and instantly shut my eyes to prevent the dizziness and spiralling incapacitation. 'Where was the horse's owner for God's sake?' I thought blindly. I guessed we must have left him miles behind during the gallop and resigned to deal with the situation myself. I would jump off.

"I'm going to jump off" I shouted to Julie as the horse finally came back on to all fours.

"Don't be ridiculous, it's fine, it barely reared at all, it's a docile as a cow. I'll tell the chap to keep hold of the reigns if you're really scared."

"No chance, he's miles back where the horse started to gallop," I screamed with my eyes again tightly shut, "I'm going to jump off before I get killed."

"He's right here," Julie informed me as the man casually walked up and took the reigns of the now static charger, "the pony hardly broke a trot, honestly, stop worrying."

I opened my eyes with some relief. "Do not let go of this monster again, I could have been killed" I barked at the grinning horseman as he led me slowly off. "And you can walk too Julie, don't go galloping off like that, you will only encourage this one."

The horseman butted in "We must go fast no? Otherwise sunset not seeing. Not much time is left for getting Sunset Point."

"I do not intend to kill myself, I would like to be around to see more than one sunset if you don't mind. I am paying. Walk slowly please."

As we ambled along the man kept looking at his watch and looking worriedly at Julie, but I repeatedly dismissed all efforts to speed up.

"I can categorically say that I hate these animals, they terrify me." I offered to Julie as we inched along the path, darkness threatening all around us. "It is like driving a car made of jelly, every time I press the clutch my foot goes through the floor and when I correct it by pressing the accelerator the clutch flies back up as if stuck to my foot. The seat moves constantly and the steering wheel moves in time with it. The wheels feel like driving on ice."

We trudged slowly on and I kept my eyes shut tight each time we veered towards the open cliff view lest I should glimpse vertigo-induced death once again. After about fifteen minutes of snail pace progress, we passed the first batch of people on their way back. They trotted past happily, bumping around on the back of the horses with glee. They were clearly not riders, but unlike me, they were enjoying themselves. A few looked uneasily at my terrified progress, I could not look them in the eye of course, but I knew this by their comments "Sunset is over, too late now." I did not reply. I was too embarrassed.

When we finally reached Sunset Point and we had indeed missed the sunset. I dismounted shakily but very quickly and told the horseman we would walk back. He protested saying 'same price, same price' concerned that we were short changing him.

"Sir, I will gladly pay you double just to take this, this, this, instrument of humiliation away from me. I never want to see it again. I am emotionally scarred."

He and Julie laughed unpityingly as I handed him the money and he mounted one horse and galloped off with the other in tow. I apologised to Julie for my ineptitude and sat forlorn on a rock to recover.

A crowd gathered around us, eager to share stories.

"Sunset very beautiful no?", "Such a lovely view I will remember for ever."

Julie nodded acerbically, "Yes, I've never seen one quite like that before."

****

Back in Pune we were still waiting for the new car and driving around in a vehicle, which though far more stylish, was in fact almost certainly more dangerous than a horse. But I was at least not afraid of it, and its repeated breakdowns enabled me to get to know Mr. Khan and his thoroughly decent approach to life very much better.

"You see Mike, a car is like a body, it must have fuel and exercise in equal measure. This car is getting too much of fuel into the carburettor and this means it keeps stopping. Too much of fuel and not enough of exercise, like a greedy Sikh man, this car is getting fat."

He also told me surreptitiously that he belonged to a local Masonic Lodge. I was surprised that a 'man with connections' would be bothering to chase around after a ridiculous fool in a ridiculous old car such as myself. I told him as much. He replied with familiar composure.

"See Mike, the thing is, it is not what you do, it is that you do it with *honour* that matters. Yes, I am having friends at the club who do not need to work. I myself choose to work. I have boys to fix cars, my job is to teach them well. With you I come in person when I can because I like you, you are a man of honour and this is important to me. You are thinking Masons are corrupt no? Wrong. Masons have honour. There is nothing more to say on this matter. When you get Indian residency permit, I forward you and my brother seconds you. You will join, I will make sure of this."

A sticker on the Herald's windscreen said 'Poona Club Member's Parking', and I asked Mr. Khan what this was. He told me that the Poona Club was an exclusive member's club started by the Britishers many years ago. Membership was difficult to come by but I should be able to go for a look round at least. This sounded fantastic, I imagined sipping pink gins on a huge shady verandah, rubbing shoulders with the town's elite and eavesdropping on conversations about cricket and elephant shoots. I made a mental note to find this relic of Albion. As it turned out I didn't have to wait long.

"Mummy is coming to visit next week and she wants to meet you and Julie," Shanti said over the phone. "I'm trying to find a place for us to go to dinner and she suggested the club, have you been to the Poona Club?"

The Poona Club was inaugurated by none other than the Aga Khan (III) at the turn of the century and unlike many other clubs of the time did not discriminate against 'prominent' Indian members. Clubs were instituted by the British with the purpose of providing convivial surroundings where they could meet, exchange pleasantries and partake of a beer or a number of pink gins after the rigours of the day's work bossing around the locals, napping, taking tiffin and getting dressed up for the club. The clubs today are a different story. While the ethos of pre-eminence remains, the clientele is likely to be a mix of haughty British sympathisers, nouveau riche businessmen and their wives with a penchant for the hoity-toity, dedicated gym goers, serious sportspeople and the sons and daughters of old money who pop in for lunch and dinner on their parent's monthly settled slate.

Julie and I arrived early and reported to the Poona Club reception, stating that we were here to dine with Mrs. Khara and others. Shanti's mother had arranged dinner for us as a distinguished member of the affiliated Delhi Gymkhana Club. The reception staff were very keen to meet her, but regretted however, that until her arrival we would not be able to enter the club. I baulked at this and insisted upon seeing the manager, who very quickly apologised and ushered us into the bar for a complimentary drink while glowering at his staff. Mrs. Khara was clearly a noteworthy guest.

Our party eventually arrived dressed in their best bib and tucker. Mrs. Khara swept in the entourage ahead of her, gliding in herself with an elegance and ascendancy that comes only from good breeding and solid self worth. The club secretary followed with slightly bowed countenance, and table waiters sprang from all angles to pull up chairs. We stood to greet Mrs. Khara as if greeting a coveted dignitary, half expecting to bow and curtsey.

"Ah, Mick and Julie I presume," Shanti's mother addressed us while brushing the train of followers into a neat line behind her with a simple wave of her pashmina, "we meet at last, I have heard so much about you, it is my great pleasure." She kissed our hands and we wilted, putty in hers.

We resumed our seats and the newcomers took theirs around the low table on the pavilion terrace and we traded greetings. Shanti, Sarasvat,

Manohar, Raman (our well appointed Sikh friend) and Vikas, Sarasvat's brother-cousin who had recently joined the company. They sat around in stiff shirts, scrubbed up like kids for Sunday school, silently looking nervous and awkward. We ordered drinks and I realised they were not at all relaxed and were not ordering anything like the normal round. In fact no one ordered alcohol and no one was smoking cigarettes. What was this? Were they afraid of Mrs. Khara? I stole away and surreptitiously asked Manohar.

"We cannot drink or smoke in front of Auntie, it would be disrespectful."

"But what about Julie and me? We are being disrespectful then?" I asked, concerned.

"No, Auntie will not mind because you are not Indian, also you are a little older than us, and you are seen as the guests so it is on your terms, you do as you wish. I don't smoke in front of my elder brother even, it would show disrespect."

"You frighten me Manohar, you really do."

We finished our (mainly soft) drinks and I took the lead and escorted Mrs. Khara into dinner, sitting her at the end of the table with Julie and I so that the others could get some respite from their formality.

After ten minutes of conversation it was clear to me that not only was Shanti's mother a well travelled woman of independence and clearly some influence, but she was damn good company too. On the subject of Shanti, I was thanked for looking after her and inspiring her at work and a whole host of other influences she seemed to think I had imparted, I told her that in fact I had done little else other than make fun of Shanti for being a pernickety prude whose affiliation to yuppiedom and squeamishness were hindering her in no small way. Shanti dismissed my common complaints as nonsense while her mother beamed a huge grin at me and, to my surprise, took on the mantle, berating Shanti for being too worried about make up to really get on in life. I whispered an apology across the table as Shanti laughed her mother's comments off. She was certainly not in awe of her mother, the others meanwhile, concealed by Shanti's leaning forward, were sneaking beer from the table and cigarettes from my packet, leaning back in their chairs to take a puff lest Auntie should discover their vice. And all of them over twenty-five years old.

Mrs. Khara continued to entertain, telling us how she felt India was losing its class, intellect and style in favour of American brashness, and that improved technology was enabling her to keep abreast of her kids better (mobile phone firmly in the table). She also told us how she had enjoyed visiting the red light district in Hamburg with her husband many years ago to see what the fuss what about, while Shanti gaped at her aghast, "Mummy, really, what will people think?"

"Be quiet Shanti, you insult us, Mick and Julie don't share your prudish nature, these are facts of life, one should know about them."

The evening came to an end and we left the club. It had been a most enjoyable experience and we vowed to take Mrs. K up on her invitation to stay with the family in Delhi sometime. On the way out we enquired about club membership and were told that the waiting list was roughly six years. Mrs. Khara intervened, "What about NRI membership? Surely something can be done for these two on a temporary basis?"

This was all that was needed, Non Resident Indians we were not, but temporary Indians we were. We were offered membership on the spot, which we took without hesitation. Yet again, being immigrants had actually elevated us to a status above that of national. I can think of no other country in the world where you actually get treated better for being foreign, and nowhere where foreign items are so vehemently cherished, I was reminded of a cartoon by the Times of India's famous Laxman (now resident in Pune as a matter of fact) where an optometrist is examining a patient and says 'Yes, there is a foreign body present in the eye, perhaps you want to keep it?'

Over time we became regular attendees at club events, rubbing shoulders with luminaries and notables from Pune society wherever possible, and dining at the club became our favourite pastime, excellent Indian staples served up by proud old waiters in embroidered tunics, who had themselves worked at the club since they were boys. But just sitting around drinking beer during the daytime was best, being sure to be between 11am and 3pm, the old British licencing hours which had not been brought into line with Britain's relaxed all day drinking repeal. We would pretend to be big shots from overseas, perhaps from one of the affiliated clubs we felt sure we'd someday visit, the Delhi

Gymkhana Club perhaps, or the Naval and Military Club in Melbourne and of course the Royal Overseas League in London, where we would definitely look in on our return.

In addition to the Club, we were able to join the British Library with full borrowing rights. As the novelty of all things Indian wore off over time, and short bouts of homesickness surfaced, we were able to slay them with nostalgic bursts of English culture borrowed from the most excellent of institutions. Perhaps the best thing about the British Library in India is that it contains nothing that is not British, and everything that is. This meant we were offered a comprehensive array of books and DVDs from all eras but from only home shores. Indian TV was beginning to grate to say the least, yes, there are a few good sports channels, and the chance to see hundreds of Indian movies at any time, but the rest of the English speaking channels (with the exception of the quite awful BBC World) are American, and show either repeats of Hollywood blockbusters starring Sandra Bullock, 'Friends' or 'The Bold and the Beautiful', which I'm sure you'll agree is not a stimulating array of entertainment in any situation. (Actually, there was a short period of around eight weeks where a Russian channel was showing soft pornography every night, quite how this escaped India's indomitable programming censors I have no idea, but it certainly sparked a wave of rapid Russian learning among the call centre staff).

Julie tried to borrow 'The Office' on DVD but was alerted to the fact that it was available on corporate loan only. She was told that it was available only to companies for training purposes. She started to explain that it was a comedy, etc, then decided to leave it, imagining the situation as a group of keen office staff are assembled in the training room and the first few bars of the signature tune are struck.

The library offered us such homely staples as 'Life is Sweet', 'Rising Damp', 'Best of The Old Grey Whistle Test', 'Butterflies', every single episode of 'The Sweeney', and perhaps the greatest and most subversive situation comedy ever made, 'The Fall and Rise of Reginald Perrin'. The latter served as a pertinent and timely reminder to me that working for a living is a dangerously fruitless pursuit, and for the two weeks it took us to sit and watch all 21 episodes I seriously contemplated leaving my clothes at the office door and just disappearing. But I didn't, and wouldn't, at least not until I got my new car, I wouldn't get far in the Ghostbusters mobile. And I didn't have enough clothes to spare either. Having been back to London briefly, I had been able to pick up a few things that were suitable for the office, but I decided to try to buy some new stuff in India to try to fit in with everyone else a little better. The Mod in my told me to sharpen my creases and set a modish example, but the coward in me told me to try to belong. So it was without fervour that I set off to town to find myself some suitable apparel.

Indian men's fashion is not for me; I might as well say that here and now. The shiny, glittery, tight fitting shirts and fake leather pants of the film stars is just not my style, and while I'd genuinely love to wear some *khadi* homespun or full length *kurta*, it really wouldn't suit me, neither for that matter would a Nehru jacket, smart though they are. So it is in the dull shops that I tried to shop. But I failed miserably, pleated trousers, baggy shirts, blazer jackets, sorry, no way. So in an effort to meet Indian fashion half way I found myself a reputable tailor and went down the bespoke route. Many people who come to India on holiday, or anywhere in Asia for that matter, avail themselves of the low-cost labour and have a suit made. I've seen them being carried home as hand luggage many times. But most of these are dodgy in the extreme. I had checked the offerings from countless tailors in Goa and found them to be very much below par. The problem is once again in the finishing; there is just always something very wrong. I heard one horror story about a West Ham United fan having a dress suit made in claret and blue, the fact that a tailor would even consider such a crime leaves me cold.

I hoped it would be different in the city and went into the biggest branch of Raymond Suitings I could find. I brought an example suit with me, a worn pinstripe from Carnaby Street with straight legged pants and a bum freezer jacket, in an attempt to demonstrate the style I wanted, but I decided to leave some of the decisions up to the tailor himself, I felt it arrogant to go in demanding without cutting him some sartorial slack. I also brought a shirt that I wanted copied exactly to get an idea of how accurate to a pattern they would be.

Upon entering the shop I was beleaguered by six smiling men behind one counter display of rolled up suit and shirt fabric. I headed to the suit material and confronted a man with prominent teeth.

"I wish to have a suit made and would like to see some fabrics please."

'Which fabric you are wanting? Which colour?"

"I think a dark grey to begin with, if I am happy with the suit I will have others made, but grey for the first to be safe."

A behind the counter melee ensued as four men scanned the racks and pulled out grey fabric, blue fabric, green fabric and black fabric at random and slammed the rolls down on the counter. One man seized each roll and threw it open for me to view. Within seconds there was a huge mess of unravelled fabric in a great pile on the counter and spilling over the sides and onto the floor.

"Wooaah, wooaah, stop, stop, please, slow down, I said *grey*, and there's really no hurry, calm down everyone please." I shouted above the commotion.

"Grey sir, grey, you like? You like this grey?" One of the men screeched, stretching across the counter thrusting a light green fabric in outstretched hands like a desperate autograph hunter.

"OK chaps, please stop all this. Put everything back apart from..." and I pointed to a number of grey fabrics among the multihued pile.

After some deliberation and having twenty or more fabrics draped suit-like over my shoulders I settled on a mid grey Prince of Wales check and asked for a price. Three and a half metres for a two piece suit would come to 3000 rupees for the material and 3500 rupees for the 'stitching', in total 6500 rupees for a made to measure suit, an absolute bargain, the same bespoke ensemble in England starts at around ten times that price. Next I selected some shirt material, a standard blue and white gingham, a plain navy blue and a fetching and rather gaudy pink stripe, at 500 rupees per shirt including stitching I could afford to make a few style *faux pas*.

I was then ushered upstairs for a fitting and produced the suit and shirt. I made it quite clear that the shirts were to be made as exact copies of my template shirt, short sleeves, tightly cut waist, medium length 'nose picker' collars. This part was easy. The suit was slightly more complex, I explained that this was not to be copied, but the style and shape were to be followed with perhaps a slightly straighter leg and slightly lower waist on the pants.

"Straighter leg than this is not possible sir, you will not fit your foot through, we are not making pants in this way sir." The master tailor insisted, looking rather perplexed. "Also lapels are very slim, you are wanting like this also? I am thinking this is very old fashioned no?"

"Sir, please do as I say on this, I am probably not your average customer, I require a suit reminiscent of a popular 1960's Italian style, it is important that you take care of the details, the ticket pocket for example, and the concealed pleats which cut the waist very tight. Please ensure these features are included, if they are not, I will not pay, but if you get this right and I am happy I will have more suits made, I assure you."

"OK sir, no problem, we are making this way, don't worry."

Being measured for a suit is a pleasing experience in any environment I am sure, and having four men doing the job at the same time, expertly shouting out measurements that are being recorded on a tiny carbon duplex pad is most gratifying. I left the shop very happy and looked forward to the first fitting in five days time.

I returned to find everything ready as promised. The shirts had been packaged which I found irritating, and told the tailor not to waste money on unnecessary presentation in future. He looked hurt. When I saw the shirts and baulked he looked even more hurt.

"These shirts are nothing like the shirt I left you, there is no cut to the waist, and my God, the buttons are awful, what on earth made you go for glitter on the buttons? They will need to be removed at once. Give me the original shirt."

I measured one up to the other, there was no comparison, they weren't even the same length. And all three were different.

"This is absolutely terrible, these will have to be made again, they are nothing like the original, why didn't you cut the waist? I specifically pointed out the waist to you."

"This way is better sir no?"

This was a red rag to a bull.

"I am paying for this for Christ's sake, I will say what is better, make these again, copy the damn shirt I gave you, stop wasting my time." I threw the shirts at him. "Show me the suit, I sincerely hope you haven't seen fit to stitch flared pants or something."

But the suit was quite different. It was excellent; a perfect fit. After grumbling a little in aftershock at the shirts, I slowly came to terms with the fact that I was wearing a beautiful suit of perfect fit. Sure, the waist needed a slight tuck and the trouser length needed adjustment, but overall it was hand in glove. I told the master tailor I was delighted and all in the shop breathed a sigh of relief.

"Yes, this is excellent, I am very happy with this." I offered by way of a truce, once again feeling stupid for blowing my top where a few direct words would have sufficed.

The final suit would be ready, along with the altered shirts in four days. I returned to find the shirts repackaged in plastic once again and my blood rose. I ripped open the packets and took the shirts to the changing room. They were good, not perfect, but good enough, all things considered. I told them I was happy. They brought the suit and I tried it on. It looked great until I realised that the trouser hem had been hand stitched by a drunken wrestler with Parkinson's disease.

"What the hell has gone wrong here?" I blurted, my habitual rage returning, pointing at the lumpy trouser bottoms.

"It is hand stitching no? Not possible as straight as machine, always like this." Offered the master tailor hopefully but cowering, addressing me through his eyebrows.

"Utter nonsense," I cried, grabbing examples of readymade pants from a display and thrusting the neatly stitched bottoms in his face, "do these look like they have been stitched by drunks? Do these?"

"OK sir, we will re-stitch and give." He tried to pacify me.

"How many bloody times do I have for come to this place? This is a complete waste of my time. Get this done quickly, today, in one hour and I will return."

"One hour is not possible sir, electricity is not there." Said the tailor, beginning to stand his ground.

I hadn't noticed that the air conditioning had gone off, this was Thursday so it would not be coming back on again, perhaps the increased stickiness had added to my rage. "What the hell do you need electricity for? You told me this is hand stitched."

"Iron is still needed sir, otherwise you will not be happy."

He had a point. I told him to phone me when it was done and stomped out without the shirts.

Two days later I called (Indians don't call you back as a rule) and was told all was ready. I entered the shop a little embarrassed given my last outburst and was intending to apologise, but as they brought the shirts out in plastic packages I almost lost it again. I but my lip and examined the suit, it was good; I tried it on and felt like a king. It had been worth the anguish and I told the tailor. He seemed happy if a little distant. I paid and left.

Wearing the suit to work the next day felt so good that I decided I needed more where this one came from. The next week I returned to huge smiles and went through the same fabric fight as before. Presenting myself in front of the tailor I bowed a peace offering and shook his hand.

"I love the suit so much I want you to make two more for me. Please make them exactly the same as the first and there will be no shouting from me. I am sorry for my conduct before, I am a difficult customer, but ultimately a good one. If these suits go well, there will be more. Oh, and one more shirt please, and please don't put it in a plastic wrapper."

"I would be very happy to do this sir. A customer like you is good for us, otherwise, how do we learn?"

You can imagine how bad that made me feel.

Over the next three months they made me five suits, seven shirts and two pairs of trousers, all excellent and all astonishingly cheap. Saville Row it is not. The fitting process involves 4-5 visits, lots of red faced shouting, stamping of feet and accusations of incompetence, but it is worth it in the end, no doubt about it.

****

With still very little to do in the office and with the onset of monsoon breathing down our necks, we decided to have another hill station jaunt to keep the spirits up. This time we chose Mahabaleshwar, another popular spot with Bombay and Puneites, some 1400 metres up and damn fresh. As it was mid May, we had to join the throng of summer visitors and queue for views along with coach party after coach party. The mountain views were again breathtaking, very Swiss in many ways, not least the piercing bright sun on the bright green of the mountain grass. India's strawberries come from Mahabaleshwar and it is easy to see why.

We stayed at the excellent Anarkali Hotel, a 1970's chrome and walnut veneer Swiss style chalet arrangement whose reception is replete with photos of its many Bollywood star customers and the rooms adorned with box-print photos of Austrian *fraulines* and scenes. A well deserved rest at great financial expense. The hotel owner and proprietor, a brash man with a protruding pate and an unmistakable air of control was most congenial while bossing us around in the way a headmaster might do at a parents' evening. He started all his sentences with 'Er, hallo-o', with a rising inflection on the O, similar to the way teenagers and Americans imply someone is stupid by saying "Hello-o? That is *so* not what I want to happen." I say they sound similar, but I am sure are in no way linked.

Two men stopped us on Malcolm Path, a thoroughfare named after British stalwart 'boy' Malcolm who founded the hill station during the 1800's. They came towards us thrusting various different leaves in our direction proclaiming,

"Blow on this and speak any languages, German, English, and large pigeon will appear."

Honestly, what is one to do? What do these people want ultimately? How is one to cope with such requests?

"I have no desire to meet with said pigeon," was all I could manage "now be gone."

They went. It was not until some time later that the ludicrous nature of the interaction even occurred to me. Perhaps we had finally gone native, numbed to and by the recurring madness surrounding us.

After three very expensive days at high season prices we headed back down to the Deccan Plain to find that we had left half of our clothes in the hotel wardrobe. I called the hotel and the owner answered.

"Er, Hallo-o? Anarkali Hotel, Hallo-o?"

"Hello, I have just spent three days at the hotel with my wife and we have left some clothes in the room." I began.

"Er Hallo-o, you are English no? Yes I know you we have found your clothes."

I was relieved, "Oh, great, that makes me very happy, how do we go about getting them back? Can you send them to me?"

"Er Hallo-o? Yes I can send them. Er Hallo-o? Shall I send them in post or by courier?" he asked in his weird way.

"By courier is best if possible."

"Er Hallo-o? This is no problem. Hallo-o? Please give me your address."

I read him my address.

"Er Hallo-o? I repeat your address is: Er Hallo-o? Shastri Nagar, Er Hallo-o? Pune Er Hallo-o? 411600 yes? Hallo-o?"

"Yes, that is correct." I was of course laughing by this stage.

"Er Hallo-o? I will send today, you pay your end OK? Hallo-o? You will get it tomorrow, hallo-o?"

"Thank you, thank you so much." I got the phone down before properly cracking up.

The next day a white linen stitched parcel sealed with red wax around the edges arrived at the door in the hands of a smiling man. 70 rupees to pay. Now that is service. Er Hallo-o? If you ever go to Mahabaleshwar, stay at the Anarkali, I will give a money back guarantee, Hallo-o?

****

Finally Mr. Khan called to say the new car was ready and that he was on his way round to deliver it. He arrived driving a masterpiece of Indian engineering, a Premier Autos Limited (India) Padmini Deluxe BE in its original shining Bayer Blue. The car is better known as the ubiquitous Bombay Taxi; there are around half a million in Bombay alone. This one too was originally from Bombay, a 1989 model made to the same template as the mid 1970's model (gear change on the steering column, bench seats, curved windscreens). And it looked magnificent. Mr. Khan had pulled out all the stops and was clearly proud of the work. I took it for a spin around the block and got used to the lurching first gear and remembered that these cars will not go into second when in motion without a crunch, for me and quite probably for many others, this crunch has become the sound synonymous with Bombay traffic lights and the taxi race that ensues each time the lights change from red to green. I returned to Mr. Khan very happy and shook his hand vehemently.

"It is wonderful my good man, late, but wonderful." I teased.

"See Mike, lateness is a concept of Britishness, this car is not late, it is delivered after the right amount of time it took to make it this good. Just remember that."

He jumped in and with a grin pressed the clutch and slammed it into reverse. Julie and I grabbed our ears as the high-pitched tune belted out.

"You better not pout, you better not cry, you better not shout, I'm telling you why. Santa Clause is coming to town…"

"Aaagh, turn it off, turn it off." We screamed as one.

He stopped the noise. 'You don't like? I thought it would remind you of your childhood."

"Please disconnect it at once," I pleaded, "You know my loathing for driving with the ears. "

"No problem, no problem." He threw open the bonnet and yanked a wire from its socket with his hands. "I remember now, no stickers, no horn, no extras, just the car. You are a strange man Mr. Mike, but I like you nonetheless."

"And you sir, I like in equal measure. Allow me to drive you home."

On the way back from his place, with the stereo blaring through the tinny speakers in true Indian style I raced the car through the dark streets with glee. The humidity made for a soaking sweaty back on the vinyl seats, but I didn't care, I had saved a dying car from the scrap yard and it felt good. As I pulled into the colony the heavy clouds finally broke and the monsoon majestically announced itself with a huge crack of thunder and a biblical downpour. We had seen a few tropical storms brought on by a cyclone in Lackshadweep in the past week and with them some flashy showers, but they were nothing like this. I sat in the car and listened to the ruthless hammering of rain on the skeletal roof and bonnet, and I laughed out loud as the rain washed a windscreen wiper clean off. Around me children and adults streamed out of the buildings to shout and dance in the rain, exactly as the cliché says they would. I phoned Julie from the car and told her to come down. As I saw her cowering in the doorway I stepped out of the car and was soaked in an instant, I grabbed her and we joined in the dancing. For a minute we were Indians, praising the coming of the life giving rain. But only for a minute - we soon remembered our awkward British composure, and, slightly embarrassed, we headed back inside.

The rain came every day thereafter. It was not the pyrotechnical display I had hoped for, and after the initial day or so it wasn't even that hard, but it came every day, and staying in became the new going out. Holed up in the colony we needed an indoor pursuit. We settled on cooking. Julie had seen an advert for Kamal's Cookery Classes pasted on the colony notice board and she rang the number and arranged for us to learn to cook Indian food properly.

Indian food is certainly my favourite. But cooking it yourself is a real no go. The worst thing any friend who is not Indian can say is "I make a mean curry, why don't you come round to dinner?" And of course you have to go, and it is horrible without exception. You are served tandoori naan from the supermarket (usually the type that is spoiled by having too many 'extras' like garlic, or God forbid, Basil. Basil Nan? I don't think so), or worse, rice (starchy as hell and boiled to death) with a gloopy or watery mess (depending on the Non-Veg or Veg ingredients) which tastes like Heinz Big Soup with curry powder sprinkled on top. Even my Indian friends don't attempt to 'knock up a Vindaloo', unless they have been so instructed by their mothers (few have).

The recent fashion for cookery programmes on TV largely ignore Indian food, with the notable exception of Keith Floyd who just throws everything he can get his hands on in a pan and produces the same meal whether he is in India, Morocco, Kenya or Dartford. The celebrity chefs do not attempt Indian food, and the Indian chefs do not become celebrities (I know I am overlooking Madhur Jaffrey, but who can actually get enthusiastic watching her blandly stirring more sugar into a syrup? This is somnolent TV at its extreme). The reason is it can't be done. If you are not Indian, you can't cook Indian food, it is as simple as that. At least that's what the conspiracy wants us to believe.

So it was with some scepticism that we embraced an attempt to overturn the theories. Kamal opened the door beaming all over her stout fifty year old frame, holding back the family Pomeranian that was barking in the shrill, pathetic way that these compulsory middle class Indian pets do. Her tiny kitchen was even smaller than ours but well organised and well equipped. We had agreed that the first dishes would be our favourites, *Palak Paneer*, Chicken Tikka, *Malai Kofte* and Chicken Vindaloo (Goan style rather than after-ten-pints-give-me-the-hottest-dish-on-the-menu style). She gave us recipes and we watched her make the dishes that we subsequently ate. It was an absolute revelation. It is no wonder no one knows how to make curry, the recipe books must be a part of the conspiracy, telling us the wrong way to do it. There are so many omissions. I am no cook, but it doesn't take an expert to note the difference in taste and texture when onions are ground rather than sliced, tomatoes are half pureed and half chopped, masalas are added at different stages and not just bundled into a paste at the start, and perhaps most importantly that the *garam masala* and garlic, ginger and green chilli paste are lovingly prepared at home and in advance. We made all four dishes ourselves the next day with implausible success. Two weeks later we had learned various dishes including the hallowed *biryani* and *makhani dal* (we have mastered neither of these yet, but we will persist) and presented them to our Indian friends for the true test. It is hard to gauge where true enjoyment stops and politeness begins but the fact that everything was eaten was taken as a good sign. The *roti* is an art that will take time to master, twenty years according to Manohar, a trained chef and terrible *roti* maker. But we had disproved the theory and it would be Indian all the way from now on.

****

As June progressed the issue of the European Football Championships arose. The tournament would be televised in its entirety right through to the final and we intended to watch every game. My workmates were equally keen and we devised a work schedule so that we could avoid the office when the games were on. Two days after the tournament started, the cable TV operators got into a dispute over revenues owed, and the two sports channels who would screen the games, ESPN and Star Sports were blocked from view for three of the four operators in Pune, ours included. Cable TV is organised on a regional franchise basis in India, leaving the viewer at the mercy of the franchisee with little course for reproach. If you have a dispute with the operator in your area and refuse to pay, no other operator will be there to step in.

Similarly, if your local supplier is lazy and doesn't fix your connection when it breaks (and it will, the cables are largely strung from trees and across buildings and are very vulnerable to both tapping and snapping) there is no one else who can fix it for you and no one else to complain to. The fact that transmission will be lost and a blue screen presented for a few minutes (and occasionally hours) several times a day means that it is almost impossible to watch a film or a sporting event uninterrupted, and nobody even complains about this, like the inevitability of a power cut, it is just accepted and dealt with (imagine the outcry if Coronation Street was continuously interrupted by blank screens, the British public would surely riot). So we did not have access and there was nothing we could do. We were missing most of the games amid terrible frustration, but we couldn't miss the England games, there had to be a way.

There was. Through painstaking detective work we isolated the area where the one supplier who had paid up their fees operated and located a hotel there with a screen. The Pride Hotel in the Shivaji Nagar area of Pune is a dull five star business hotel with, as things turned out, probably the best sports bar in town. Our original plan of hiring a room and using it just to watch the TV was not necessary, and despite the ten kilometre drive each way, the place stood us in good stead for the tournament. As the competition progressed and England progressed with it, and as word got round about the hotel, the crowds multiplied and intensified. The extent to which British companies were operating in Pune became shockingly apparent, as literally hundreds of partisan English clients and workers escaped the offices for the camaraderie of the Pride Hotel for two brief hours of football. They brought flags, horns and replica shirts, and they brought a riotous noise to rival Eden Gardens in full test match voice (ok, maybe not quite).

On the day that England were finally dumped by the host nation Portugal, the crowd was split 40:60 between a strong Goan Portuguese following and a stronger native English one. For one night only, the Pride Hotel became every other ex-pat sports bar in Hell. The chanting Neanderthal mass of drunken English Oafs displayed an astonishing lack of poise that the Portugal fans tried heroically to match, but there was never any hope, it was a one horse race from the start. There was no violence despite the result, but it was very, very close. I advised two celebrating Goan Portugal fans to do their celebrating elsewhere as I overheard two inebriated Mancunians discussing after match tactics in the toilets. And the mess. Lets just say I wasn't too proud to be English.

On the way out of the hotel bar, the manager thanked me for my custom. I looked back into the disarray where tables and chairs were upside down, beer glasses and bottles were scattered and two lonely chumps were sat with their George Cross heads in their hands, and I could find no reply.

Chapter Fourteen

While the office had promised to get busier, there were further delays, and the persistent rain fuelled my boredom as June came to its end. Maybe it was the rain or maybe it was the boredom at work, but my already thin on the ground patience with life's little quandaries was reaching its limit. I only went to work on alternate days, and when I did, if the car park was full I would turn round and come back again. When possible I took to parking in the VIP car park in which I was apparently not entitled to park. The security guards would get mightily upset at me parking there and do their utmost to try to stop me, insisting on seeing my ID card to ascertain my status. I would either refuse and act like a madman or simply say "I'm sorry, I don't speak English, I haven't a clue what you're talking about", my irascible conduct fuelled by ennui. I couldn't wait to be confronted and contested by someone who actually was entitled to park there so that I could explain that laws were there to be flouted and vent my spleen and contempt further on their authority. Actually my 'common man' car was most conspicuous among its Tata, Toyota and Honda counterparts and I think it was this that irked the security guards most. And so it was with further distaste that I conceded that yes, India had finally succumbed to brand snobbery too.

Julie embarked upon a short art project to produce paintings of the things in her life that didn't work. I was at the top of her list. Below me were: the TV, the electricity, her mobile phone, our laptop, the lift which had been off for two weeks (denying us Theme From Love Story) and our cleaner who, when she did show up, showed no initiative towards anything that was dirty and defied her job title entirely. Everything that gave Julie communication with the outside world was broken, and the metaphor was very ominous. We needed another break, and despite the rain, we braved a drive to Lonavala and Khandala, some 70 Kilometres away. We had been told that both these hill stations were largely just viewing points for coach parties, but we thought they would make a good base at which to stay while looking at the caves of Karla and Bhaja, two clusters of early Buddhist rock temples famed for the ancient art with which they are adorned. More than anything we wanted to get away and test the car to see how it fared on a longer journey, we were, after all, intending to drive it back to Goa at some point.

We had both seen the Khandala views from the Mumbai-Pune Expressway on various occasions. The highway is lofted upon a vast flyover right through the centre of Khandala, which offers fabulous views for the expressway traveller, but terrible ones for the Khandala dweller, or at least the Khandala dweller who is not beguiled by the beauty of colossal concrete columns at very close quarters. We took Mr. Khan's advice and avoided the expressway in favour of the old Mumbai-Pune road. Although it was bound to be pothole ridden as a result of the rains, Mr. Khan reasoned that it would be much slower and safer than the highway.

"Mike, please do not use this expressway, it is not safe. Indians do not have the discipline to drive on such roads, overtaking from the inside and BAM you are dead. Take the old road it is very nice, you can stop to take snaps of waterfall. On expressway you cannot urinate also."

It was mid July and the rains had abated and were now down to regular showers. We set out on the Friday morning while Mr. Khan would be at the mosque praying for us. The weather was clear, but the roads were far from it. After an hour's driving we had barely reached the edge of the city. The roads were an abomination. I had heard how the rain tore the roads apart each year, and had experienced how driving to work down a rough track had become proper off-roading, but the holes in the city's main roads were astonishing; I feared we may disappear. The traffic was so heavy and there were so many minor accidents caused by two-wheelers swerving at the last minute to avoid unexpected death pits that we very nearly turned back. It was two hours before we were clear of traffic and although we were far from clear of holes, the road became acceptably empty. I noted how indifferent we had become to the rigours of the Indian road, a year ago this would have been a story of near death and fear, but today it is just another hassle of Indian life.

We drove through lush green countryside, English countryside perhaps. I had never seen India at this time of year and it was majestic. The mountains were awash with green and alive with streams. We followed the road as it meandered over hill and dale parting an occasional forest of verdant trees, their leaves throbbing with new chartreuse life.

We reached Lonavala after three hours of continuous on-road off-roading and the car was panting a little but was at least in one piece. As we drove the further eight kilometres to Khandala however, it seemed to sense we had arrived and began spluttering and kangarooing treacherously. We pulled nervously into three or four hotels in search of a room with a view, but without luck. We knew we would have to stop soon and that if we didn't, the car would. But we stubbornly had to have that room with a view so we doubled back to try some places we'd overlooked and in doing so, reversed into a three point turn on an incline with a cliff edge blind corner to the left. The car spluttered to a terrifying halt. It wouldn't start. We were stuck at the exact point in the thoroughfare where bus and lorry drivers see the tempting downhill stretching out carpet-like before them and pull out of gear to start leisurely freewheeling, the problem was that we were in the middle of that same road, concealed by a stack of granite the size of your average Sainsbury's. Julie tried bump starting in reverse to no avail. The only option was to free wheel backwards down the mountain road until a layby of some description materialised. It did not, but traffic did. As we meandered backwards we met a bus climbing with horns blaring in disbelief as the rear of our little biscuit tin car played involuntary chicken with its front end. It swerved to miss us at the last minute, risking the lives of all its passengers and passed with inches to spare. If a vehicle had been coming the other way I would have considered us responsible for genocide. (Can it be deemed genocide if it is down to unintentional stupidity?). We continued to career backwards down the terrorising mountain road as Julie tried in vain to bump start the little car in reverse. Thank God she was driving and not me because realising this plan was fundamentally floored, she had the foresight to spin the car round and jam it into second gear, firing up the engine and wantonly careering the vehicle across the road to a waiting layby some hundred metres on the opposite. I put my eyes up to the heavens as we stopped. Once we'd got our breath back I promised Julie that we'd get a decent car when we finally returned to England, this had been one breakdown too many. I got out to clean the jets and the carburettor, which I knew would be full of gunk and the source of the car's jumpiness. As I lifted the oil filter off, I slipped and upended it, depositing dirty oil sediment all over the engine and my forearms. At the same moment, two men materialised to investigate the white people in trouble at closer range. They were not mechanics and I was not polite to them.

With clean jets, the car was happy again and we eventually found the Dukes Retreat Hotel and checked in. The hotel is famous in these parts and one can expect to rub shoulders with Bollywood stars. The room tariff was extortionate and there were no stars. No stars, that is, apart from the grey hornbill that sat staring at us in a tree adjacent to the balcony of our room. I had never seen one before. It was not the colourful great pied hornbill that is so famous, but its shabby poor cousin of a lower caste, perhaps even more magnificent in my eyes, give me a scruffy sparrow in preference to a fanning peacock any day. Seeing the bird made the whole trip worthwhile, which was a good job, because it was hard to find anything good to say about Khandala or its bragging amusement arcade neighbour, Lonavala. That evening the rain came on heavy again and didn't stop for two days. Lounging around the hotel and staring at the grey mist for hours took its toll, there is after all only so much Kingfisher beer one can drink, and we set off back to Pune early. We went via the caves that had brought us on this death-defying trek to find them closed. No explanation, no information, not even any misinformation, just closed. We drove back in silence and went straight to bed.

****

Bored. Bored. Bored. It was a surprise, but we were bored out of our minds. We couldn't find anything new to do. After a year in India, things had become commonplace. The hilarity of situations had passed and things became ordinary to us. The wonder had gone. The honeymoon period was over, and we were bored. I'm sure if we felt we had some purpose, some value, things would have been different, but we didn't, and it felt dull. We would wake up some days, look at each other, shrug and go back to sleep. The newspapers seemed to carry no news, and the TV was just stultifying, so poor that we began to look forward to 'Pet Star', a programme where dogs catch Frisbees and parrots make noises after which Americans clap.

Finally the hiring for the new project started and I was given something to do. I had not been overly impressed with the hiring to date and intended to improve it. The main problem seemed to be that the people selected had both good English and intelligence. This might seem to be a prerequisite, but intelligence is not all, and to have a flair for the drudge that is telephone interviewing is far more important, indeed intellect is often a hindrance.

The task was to hire ten full-time and ten part-time employees to work on an 'omnibus' survey, a multi-client questionnaire that changes every day. The clients buy questions and we ask them of people and supply their answers. The idea is to give a quick and cheap picture of a market, maybe to ask how the public feels about an advert or a product or service in a very general sense, without going into specifics. This means that questions might be about anything, and the switch from one subject to another can be very abrupt. PR companies use this method a great deal, so questionnaires are likely to include magazine polls such as 'Which actor do you fancy most?' or 'Have you ever induced vomiting after meals?' The diligence and composure necessary to deliver a set of questions such as, "Would you expect to have sex on a first date?" followed by "Which do you prefer, bananas or cucumbers?" and culminating in "Does your vacuum cleaner have a bag to collect rubbish or does it deposit straight into the drum?" is less a matter of intelligence or intellect and more a matter of brass balls.

With this in mind I went to meet the first candidates who had already been whittled down from the two hundred or so who had arrived for the 'walk in interview' having seen an advert in the paper promising huge salaries for very little work. They had been quickly assessed for aptitude and ability to speak English before going in front of the human resources department for short listing. The teams who have a requirement for staff would then see them. At no time does anyone tell them what the job entails other than whether it is 'voice' or 'data' meaning they will have to speak on the phone all day or punch numbers into a computer all day. I found this shocking. I have never been for a job interview without first researching something on the prospective employer and certainly not before having a very good idea of what the job I'm applying for is. Imagine:

"We are happy to inform you Mr. Sheridan that you have been chosen as the successful candidate, congratulations, you start tomorrow."

"Great, I'm delighted, may I ask what the job entails?"

"Yes, you will be required to spend eight hours a day pressing the numbers 1 or 0 on a computer keyboard."

Sarasvat and I sat in a glass walled interviewing room as the candidates were paraded past to the 'holding area' to await their turn. I had no idea what to expect, having asked Sarasvat to let me see how he normally went about things before offering any improvements of my own. I had also told him I would remain largely silent throughout the interview, as it would already be intimidating enough for them to have an unexpected white man listening to their efforts to sell themselves, without me interrupting.

Sarasvat had been given the application forms and *résumés* and selecting from the top of the pile called out the name 'Jokim' without giving the papers a second glance. As the man stepped forward Sarasvat could see I was looking surprised and he leant to me and said "No point reading it, it's all a load of bullshit anyway." Jokim came in and stood bolt upright in front of the desk. Sarasvat gave him a cursory glance up and down and said, "please, sit, sit." Jokim sat rigid, staring straight ahead.

"Tell us something about yourself Jokim." Sarasvat demanded.

Jokim took a deep breath, "My name is Jokim Tirkey, I am nineteen years old, I have done my schooling from Kataria High School and am in second year of Bcom from Fergusson College. My father is engineer, mummy is at home only. My interests are sports and reading. That's it."

Sarasvat and I looked at each other, Sarasvat urged Jokim to continue, "tell us a little more Jokim…and try to speak normally, try not so sound as though you're reading a script."

"I'm sorry sir, I didn't get you…"

"Tell us more…" snapped Sarasvat as I covered my face with my hand, looking away.

"Err, I am Jokim Tirkey, I am from Pune itself. Err, what else do you want to know?" he seemed genuinely surprised that his rehearsed synopsis was insufficient.

"OK," Sarasvat looked strained, "Let 's try something else, we'd like you to talk for one minute on a subject so that we can get an idea of how well you speak English. Choose from either 'the effect of television on young people' or 'the person you admire the most', take some time to work out what you want to say and start when you're ready…take your time and relax, we want you to speak for at least one minute."

Jokim looked horrified while Sarasvat and I looked at the ceiling. Realising he couldn't get out of this, he took a deep breath and began. "OK, I'll talk about the person I admire the most OK? The person I admire the most in the whole world is Mother Teresa. She is an amazing person because she helped so many of the others but not for money just because they need it. They are sitting there in the street only and she is coming to help every time. That is why I think she is the best of people that I admire. That's it."

"Well, that wasn't one minute, but that will do, thank you Jokim, if you just wait outside we'll let you know." Sarasvat was smiling a big false grin while writing 'reject' in pencil on poor Jokim's resume. "Please ask the next person to come in."

As our disenchanted candidate left the room Sarasvat shook his head with despair, "Why do they send these people to me? This man could not speak English at all. I'm sorry to waste your time."

"Not at all, I take it he wasn't up to scratch then?"

A timid looking girl who appeared to be about twelve but was in fact eighteen took us through a similarly prepped oration and a similarly inexpert address relating to Mother Teresa's virtues before handing over to the third candidate, another mouse-like teenager who repeated the same act almost verbatim. As she left I expressed my surprise at Mother Teresa's altruistic popularity, surely not the choice for aligning oneself with the brave new world of global business?

"They always choose her. I'm sure the careers advisor tells them to. Think about it, who else is there? Gandhi is too political, particularly for call centres, and Tendulkar is too successful. Humility is the attribute everyone thinks employers are looking for, they're wrong of course. But I'd rather they pick Mother Teresa than their own mother, which is the next most popular choice."

"Who would you pick if you had to go through this?" I enquired of Sarasvat.

"I don't know, I had someone talk about Inzamam-ul-Haq once, the Pakistani cricket captain. That was brave. I employed him. I'd probably choose someone in business, Tata maybe, JRD Tata that is, you know him?" Sarasvat petitioned.

"No, but I would guess he's something to do with the Tata business, why don't we ask the next candidate if he knows who he is?" I teased.

As Ravi, a tubby, drained looking boy walked in and stood terrified before our desk, Sarasvat obliged me.

"Sit down Ravi, and before you tell us anything about yourself, do you know who JRD Tata is?"

"I'm sorry sir, I didn't get you...?"

"Never mind, tell us about yourself." Sarasvat smiled secretly to me and I did my best to conceal the beginnings of the giggles.

Ravi stood up like a soldier and spluttered his nervous torrent, "My name is Ravi, I am from Pune itself and I am living with my family..."

"Ravi, Ravi, calm down please, and sit down please. Be relaxed and tell us about yourself. Take a deep breath and take your time." Sarasvat could be gentle when he wanted, but his smile was broadening and the silliness was rising. Ravi sat wide-eyed and continued at speed.

"...I am living with my family, my father is a teacher and my mother is a home wife and my brother is from fourth grade and my sister is no more."

I bit into my shirtsleeve as Ravi continued getting louder and louder...

"I have done my schooling from Dastur and I am studying for my Bcom at University of Pune. My interests are reading and I love to ride my bike. That's it." As he finished his speech, Ravi's eyes glazed a little as if his obligations had been fulfilled and he could retreat back into his shyness.

Sarasvat just laughed out loud. "OK Ravi, that was most entertaining. We need to assess your spoken English so we'd like you to talk for one minute on a topic that we choose for you, will you be ok with that?"

"What topic is that sir?" Ravi's eyes once again gleamed terror.

"We'd like you to choose the person you admire most and tell us why you admire them."

Ravi wasted no time. He stood, and remembering, sat again, looking straight ahead. "This person I admire is my mother because she cooks for me my food and my father's and she is an excellent female. That's it."

I couldn't hold it any longer and blurted a huge laugh, which I think I successfully disguised as a cough, but couldn't be sure.

"So Ravi," Sarasvat went on "You love to ride your bike and you admire your mother because she is an excellent female?"

"Yes sir."

"And you have nothing else to say?"

"No sir."

"OK, well thank you Ravi, that was most enlightening. Please wait outside with the others."

As Ravi stood, bowed and turned, I did everything in my power to suppress the cackle, but it came blasting out as it always does. Sarasvat joined me spitting laughter all over the desk and I felt like a school bully, humiliating kids for fun. It was awful, just awful. It took us nearly five minutes to recover sufficiently to beckon in the next applicant. As we pulled ourselves together, we resolved to try one more before calling a halt to the charade. The people we were interviewing were totally unsuitable for the job, we hadn't even got past the preliminary introductions before realising how inappropriate they were. We hadn't even asked them why they wanted the job.

Before we had fully recovered, Shreya strode into the room, sat down and placed her motorcycle helmet confidently on the desk. Sarasvat and I rearranged our shoulders and introduced ourselves.

"OK, Shreya, tell us something about yourself."

"Sure, but before I do, can I just ask a few quick questions to make sure we're not wasting each other's time?" she asked, surprising us somewhat.

"Err, yes, please do."

"Can you tell me something about the job I'm interviewing for? They haven't really told me anything about it. Will I be making calls or receiving calls? What am I supposed to do exactly?"

Sarasvat took some time and filled her in on the mundane horrors of the position and she seemed genuinely pleased.

"OK, that sounds good," she told us, "I'm only interested in doing something which is fairly tough and challenging, I really wasn't interested in doing any data work, and I told the HR that I only wanted to do outbound calls because I've done customer service before and it's so boring sitting there waiting for someone to call, its ok when they do call because they're nearly always slightly angry and complaining and this makes it quite fun to deal with..." She looked up at our astonished faces, "What's the matter, have I said too much? I was just being honest about my experiences."

"No, no, quite the contrary, you seem to be exactly the sort of person we're looking for, in fact..." Sarasvat looked at me and I nodded, "Shreya, you've got the job, when can you start?"

The recruitment followed a similar pattern all week, and after some further consultation with the HR department regarding the unsuitability of protected middle class kids with an indifferent outlook on earning, or shy, sweet kids who just didn't stand a chance, Sarasvat and I were able to select an average of one person from every seven interviewed. They were not all quite as forthright as Shreya, but we assembled a group of suitably temerarious contestants nonetheless. Nearly all of them were one-time college dropouts with some experience of the unforgiving immediacy of survival within the Indian social system. As the apparatus of new employee bureaucracy was set in motion pandemonium broke loose. Apparently, Sarasvat and I had offered too much money to the undergraduates, the assumption being that people who have not successfully completed a degree are worth less than people who have, even though the task they are hired to perform is identical. It also transpired that we had contravened the company policy by even offering positions to undergraduates at all, an email I received from the head of HR stated that to employ undergraduates would 'dilute the high calibre the work force'.

I stormed into the man's office and questioned his policy. He was unbending. I asked him to at least consider my point of view, indicating that his policy was more than a little elitist. He did not flinch. I told him that if we did not start these people in training in the next three days, the client would take their business elsewhere. He told me this was not his problem and that he could not be made responsible for this. I asked him how he intended to maintain his position in a support function if he prevented the company from operating a business for him to support. He looked away in silence. I told him that if he did not make offer letters to the undergraduates in question, I would resign and cite his pompous attitude as my reason for leaving to my boss, who would come looking for his head. The next morning the offer letters were on my desk. That spineless little shit.

The next stage of preparation would be the voice and accent training. This is a well-honed, professional course that all agents are required to take to assess their ability to be understood by their target audience. It involves voice and accent neutralisation to overcome their MTI (Mother Tongue Influence), tone, speed and enunciation, essentially de-Indianising them in order to be understood. It is hard to deny that this is the ultimate humiliation for Indian workers, they are effectively disguising their identity because their identity as Indians is simply not good enough. As Indians, people will not talk to them, or at least this is the theory, and this is why many International companies are still reluctant to capitalise on the Indian workforce. The truth is however, often quite different, at least it is in the realms of market research. A percentage of people in the UK will put the phone down on Indian callers because they don't understand them (or won't understand them - it's a convenient excuse to get out of spending time answering questions), and a small percentage will hang up because they simply won't talk to foreigners (yes, these people still exist), but most are quite happy to talk to anyone who is polite, professional and clear, and this is exactly what is taught in voice and accent training; clarity and politeness. Indian English is often spoken at breakneck speed and can be very difficult to follow, far from clear to the English ear, and as for politeness, the stereotype of the obsequious Indian embroiled in English language pomposity is entirely absent when talking on the telephone; you are lucky if you even get an acknowledgement that the call is over, let alone a 'goodbye'. 'Minimising the MTI' is now apparently also going on in call centres in Glasgow. The tone, speed and enunciation of English spoken by our new recruits would not disguise their Indianness (pretty pointless considering at the outset of each call they are required to state their real name and that they are calling from TDS, Indian Office), but it might help prevent their Indianness hindering their chances of success.

UK culture training goes alongside voice and accent, and not being much of a Henry Higgins, this was something I could help out with. The first session I attended was hilarious. The trainer was following a syllabus containing the most ridiculous information about Britain imaginable. The new recruits were learning such essentials as how many digits there are on a standard British credit card, the lengthy procedure for acquiring a British driving licence and an in depth discussion of road markings and their relation to laws governing the parking of motorised vehicles. I elected to intervene, but really didn't know where to start. I thought long and hard about how best to provide an Observer's Guide to British Culture that would be useful to an Indian call centre worker. How does one attempt to condense such a vast tome into a useful three-hour session that would be remembered? Eventually it came to me and my agenda for the session eventually contained only four points:

Food and Drink
Sense of Humour
Slang
Profanity

I imagine that the resulting session will not be forgotten by any who attended. Defining the terms 'Air Biscuit' and 'All Mouth And No Trousers' to twenty smart Indian teenagers was an experience I will continue to laugh about for the rest of my life. So it was in buoyant mood and with great team spirit that the first day of work for the new team approached. But we waited and waited as things were continually postponed, and with typical TDS bathos, that day never came. Having assembled an exceptional council of undergraduates and tutored them in the ways of righteousness, the client delayed again and we were back where we started.

This was it for me. In giving up my short-lived freedom for this job I had anticipated a blend of work success and work failure; a mix of highs and lows with the potential reward of business *éclat*. What I got was frustrated semi-failure, the corporate equivalent of counterfeit drugs - a moment of psychosomatic euphoria followed by a crushing realisation that I'd been conned. I felt for the first time the reality of Work. A prickly line of sweat covered my forehead as the realisation hit me that I had been wasting my time. There was to be no reward. The security of a career was sham. Work is uncomfortable and boring and unrewarding and futile. There are no real highs, just money and disappointment. But as my anger grew, a strange elation grew alongside it. It dawned on me for the first time that I really didn't want to do it anymore, I genuinely had no interest in carrying on with the charade that work was good for me. An enormous sense of well being filled me as I realised it was over. I would quit. And I would never go back.

I told Julie. She smiled. I fixed a meeting with the boss.

'You know what you can do with your job? You can take it, and your eye teeth, and your franking machine, and all that other rubbish I have to go about with, and you can shove it right up your arse '.

But again, I didn't say that. I didn't even want to. I told Avinash that I saw little point in continuing things as my work was all but done. While there were clearly not the one hundred or so staff we had anticipated by this stage, I had spent so much time conducting faux training sessions to fill the working day that I had pretty much taught the existing staff everything I knew. I told him I would go back to Goa and be available for consultation should this be required, and that if work did really start hotting up I was prepared to come back for short periods on the same terms as before.

And of course my resignation was quickly and professionally accepted with a well wishing handshake. No imploring or beseeching, not even any real expression of sorrow, just another businesslike anticlimax.

Chapter Fifteen

Leaving the job was insufferable. I had reluctantly agreed to stay another six weeks to 'finish things off', whatever the hell that meant. It transpired that waiting is what it meant, waiting until September before we could leave. I have explained that I am no Siddhartha and do not comfortably have 'the ability to wait'. So I became even more frustrated and angry about my inability to escape. Had I no control over my own destiny? No. I am owned by Work. Work sets the rules and I obey. But I didn't work there any more. No matter. They still dictate. They say 'wait' and I say 'how long?'

Cornered, I waited in the shadows like a fiend. Everything irritated me, and everything felt my wrath. The IT department got it first; my computer broke again.

"Rahul? Mick Sheridan. My laptop is playing up again, can you come over and have a look?"

"I'll send someone," replied Rahul nonchalantly.

"No you won't, you'll come yourself. You have two minutes. I will be holding the computer out the window, if you don't come I'll drop it."

He came. I brought the computer back in the window to safety. He looked terrified. He couldn't find anything wrong. He tried to leave.

"What's your policy on deliberate breakages?" I asked cruelly, "I mean, what would you do if I knocked the computer off the table accidentally on purpose? Would you just get me another one? What's the warranty situation? Is it covered?"

"I'll send my boss."

The boss came and I gave him the same disrespectful line of inquiry. He didn't answer; he just left. I got an email from my boss relating to an 'IT incident' that I had been involved in. I replied saying "I hate working here." He didn't respond further.

The finance department found it impossible to believe I was leaving after so short a period, and their bureaucratic wheels were as difficult to stop as they were to set in motion. I had been paying a provident fund, a compulsory government pension, and I was determined to get it back. It would take six weeks just to get the form. Apparently it would take six months to a year to get the money back.

In mid September I was informed that with four weeks left of my contract, I would still be required to attend an 'offsite' meeting in Mumbai. I was furious. Back in March I had attended with such enthusiasm, and a mere six months later the idea was abhorrent. The prospect of so many hours of disinterested clapping while indoctrinated bromides collected glass obelisks engraved with their mainstream achievements while continuously thinking 'Don't clap, throw money' rankled to say the least. Maybe this meant I had finally made the break from the mollification that is work? After some consideration I decided that this idea in itself excited me enough to warrant an attendance. I felt if I could remain disinterested throughout, I might be able to ensure this was my last 'offsite' ever.

When I was informed that I wouldn't be flying but would be driven by minibus along with my workmates, I really didn't mind. The expressway is a 'good' road and it only takes around three hours from Pune to the Mumbai office in uptown Powai, and besides, I enjoy the contrast of the Ghats and the City, passing from luscious green to industrial hell, an ironic road to civilisation. The day prior to departure, I was informed that it would be a 6a.m. start. The morning flight was confirmed full, so it was in solemn displeasure that I rose at 5.30 and headed to the office.

As I arrived at 5.45 no one was there. There were apparently over one hundred people due to attend and not one of them was yet apparent. I sat on a chair by the chai stall and allowed my blood to come slowly to the boil. I guessed some people would be inside the building having entered from the rear staff door and I could imagine their buoyant mood as I scowled in silence. I also guessed that the majority would be late, my phone rang and Sarasvat confirmed my speculation, himself and Shanti would be there around 6.30, could I ask the bus to wait for them? I checked the time, 6.10, the bus still hadn't arrived, I ended the call without a response.

The buses came at 6.30. The people came at 7. We left at 7.30. I used expletives. I knew what would happen - the irrepressible Indian sense of fun would triumph over my anger and they would sing for the entire journey, but I wasn't going down without a fight. As each grinning fool entered the bus bushy tailed, I glared hatred into its strident soul. The engine fired, the music started and I rose to stop it.

"Turn if off now for fuck's sake."

He couldn't hear me above the distortion.

"If you don't turn it off now I will fucking kill you." I screamed as those behind me started to sing along, "And you lot can fucking shut up as well," I bawled at them in loud isolation as the music abruptly stopped.

I sat back down with knitted brow and silence reigned for two full minutes before the shock turned to murmuring discontent. Sarasvat was approached to speak to me but declined sensibly. Finally I was approached by one Inrad, a senior manager who outranked me considerably, a man of charm, wit and good nature, a man of the people and a man of the company, a well liked, well likeable chap.

"Mick, these guys want to play some music, this is something of a day out for them and they want to enjoy it…"

"Yes, fuck off."

"Right, we're having music and that's it, I don't appreciate being spoken to in this way either…"

"Yes, fuck yourself. Do what you want you fucking corporate prick."

The music boomed for the entire journey, and while they all sang and had fun I wound myself into a tighter and tighter coil. The driver didn't have a clue about the directions and neither did anyone else - four and a half hours of brooding sulk exploded as we reached the conference hotel some six and a half hours after I had got out of bed. Apparently I had been booked into another hotel and as the bus was so late, was now expected to go straight into a hall to listen to back patting approbations and smile along…I located Harsha who had organised the event and made my position clear to him.

"First I am told I can't fly despite what it clearly states in my contract, and that I have to get a fucking bus with all these fucking fools who never do anything on time. We leave hours late and they sing fucking songs like drunken idiots all the way. The driver is so fucking hopeless he doesn't know the way, and this lot are so fucking backward that they don't either. I arrive three and a half hours late and am now told I am not staying at this hotel. If you think I am going straight into a business session after that then you are fucking wrong. Get me a taxi to my hotel right now, tell it to wait until I am ready and only then will I return. I am also booking myself a flight back. Get me a taxi now or I will become violent."

As I fumed in the back of the taxi I realised I had forgotten my passport. I was staying at Rodas Hotel in Powai, a hotel I had used many times, where the staff are very accommodating, maybe they'd be ok about the passport, perhaps it wasn't the life or death issue it seemed to be everywhere else?

"We cannot let you stay without your passport sir." The front desk monkey informed me with a smile.

I put my head in my hands. He continued, "You see the thing is…"

"SHUT UP. SHUT UP YOU STUPID MAN. GET THE HOTEL MANAGER OUT HERE NOW. AND DON'T SAY ANOTHER WORD TO ME."

I sat on the adjacent sofa muttering semi-incoherent invective along the lines of "fucking hate India…move to Pakistan…people civilised there…"

After a thirty-minute argument during which I gave a false passport number and was caught (probably not a good idea to give 12345678 really) and Julie was finally called to read the number from my passport over the telephone, I got a room. As I sat on the bed drinking the second beer from the minibar I reflected as to whether I had been in India too long. The plan was to stay another six months, and although in Goa there would not be any corporate frustrations, the inherent perplexities of day-to-day interactions would not go away. Perhaps after a year I had really had enough? Why labour the point? I had failed to quit working and failed to do anything worthwhile while working. I had failed to let go of my childish, hot-tempered 'Britishness', and worst of all, I had failed to learn anything of India, I was no wiser than the day I set out, just a whole lot angrier.

I dressed and returned to the meeting, but I really didn't have the stomach for it. I had already proved to myself that I wanted nothing more to do with it, but it was as much my own attitude that I detested as the 'corporate line'. I apologised to those I had offended and headed back to the hotel. I booked a flight back for the next day. I would be expected to attend the management meeting the next day but I really couldn't face it, what would be the point? Sarasvat called me to ask if I was coming to the evening 'do', I declined, saying that I was too tired and angry. Instead I went out alone to get drunk.

I sat in the piano bar in Hiranandani Gardens opposite the hotel and drank beer after beer, surveying the contrived pastiche pub with distasteful eyes. The two mini-skirted women swaying either side of their multi-bearded keyboard hero, singing 'Hotel California' in nasally whines failed to raise even a laugh

(I'd been counting 'Hotel California's' since we'd arrived in India and was up to number fifty-eight). The predictable non-discerning yuppie clientele sat groomed and bored like the plastic gangster heroes in their films. There was nothing for me here, nothing at all. I ordered more beer to drown out the din of my disquiet and noticed with surprise that a man at the end of the bar was reading a book. On my way to the toilet, I interrupted him to ask what he was reading. From the look of him I expected Alvin Toffler or some other business guru keen on motivational statements. What I got was most unexpected, as he flashed the garish cover at me I read 'Salman Rushdie - The Ground Beneath Her Feet'. If I had been sat on a stool I would have fallen off it, and I told him so.

"You're the first Indian I've met in a whole year who actually reads quality literature, well the first under fifty years old anyway." I told him.

"You haven't been paying attention then." He retorted with a smile.

Puneet worked for a call centre in Powai and was twenty-three years old. We discussed Rushdie, Graham Greene and Peter Cook. He knew everything about them. After another drink we were reciting lines from Peter Cook's sketches. I asked him where his contemporaries were, where his likeminded peers were, he told me they were at work or at home, but ensured me that they were numerous. He denied my notion that the Indian yuppie-set were not discerning in their taste in entertainment, preferring to see them as individuals with a taste for everything new, undeveloped in a Western sense perhaps, but very much more developed in an Eastern sense. Back at the hotel, I mused on his point and became further embarrassed about my previous behaviour and presumptions. 'Staying on' in India for me would necessitate a much more openly

'Eastern Sense' if I was to engage more and learn anything about the country and how to live in it. And I knew what this 'Eastern Sense' was. It was patience, nothing more, nothing less.

Shortly after arriving back in Pune on the flight next day I got a call from Avinash asking where I was. Surprised that I had returned, he told me that the upper management had been looking forward to meeting me again, and that they wanted to offer me the opportunity to keep working for the company on a part-time basis while in Goa, same salary, less work. I told him I didn't want to know.

"Mick, I know you've had enough of being at the office, but this way you could help finish what you've started while only working something like one day a week. Listen, there's a meeting in Delhi next week, come over and meet a few people, there's a whole different group assembling to run the new Gurgaon office, a much more strategic approach which you'll enjoy. At least come up to Delhi, what have you got to lose?"

I told Julie and she threw a load of cushions at me.

"You're going to accept aren't you? And after all this? You are absolutely amazing..." she admonished.

"Look, I'm thinking about the money. This will pay for another six months in Goa and I don't have to do anything in return. This will be easy, you'll see. Please don't worry..."

On the plane to Delhi the pangs of guilt made me feel sick. I knew that if these people flattered me I would do whatever they wanted. My anger had just been vain and misdirected, and my vanity was so alarmingly unprotected that their attentions would effortlessly slay my notions of escape. I would have to be tightly vigilant. I would stick to my guns and fight to continue working on an entirely superficial basis. I would accept an offer but not honour it. For the first time in my life I would do nothing but draw the salary. I would hang around in Goa, get paid, and would work only very occasionally and very slowly, they would have to employ a little of the aforementioned 'Eastern Sense' with me...

I needn't have worried. With typical TDS form, the 'different group' who were assembled comprised a few confused new employees, myself and a few relocated old boys who were equally unsure as to why they were here. Avinash explained that the three expected bigwigs were 'otherwise detained'. While people introduced themselves and their strategies I appraised the Gurgaon skyline from fourteen floors up. The 'village' of Gurgaon has become the unwitting business capital of outsourced India. The secretive glass towers of the world's largest companies that line the highway have annexed Delhi as a sort of scruffy London Docklands. Docklands, that is, with the charm of Milton Keynes. In an attempt to create a convenient working environment for all, Gurgaon town planners have re-created Slough.

When my turn came to speak I laid things out very clearly. I didn't want to get this wrong so I read it from the pad on which I'd written it while the others were speaking.

"I am very unsure as to the genuine possibilities for successful outsourcing in my field. Over the past year a mix of client indecision, political correctness, operational apathy and Indian employment law have served to hinder our project enormously. I am unable to predict the future, but at present there is nothing further for me to do. With these factors in mind I have decided to relocate to Goa on full pay where I will be available for consultancy. If client interest intensifies, I will ensure that their intensions are met, if it does not I will consider my position again after a six month period."

And with no one there to contest this, my position was sealed. Now all I had to do was stick to my work-but-don't-work plan.

As this was my first visit to Delhi I got away early to see some sights. Lutyen's Delhi reminded me somehow of Paris and the familiar Chandni Chowk reminded me of Bombay, but the scale and cleanliness of Delhi surprised me. The Red Fort was most remarkable for its quiet, and the Qutub Minar for its grand nonchalance. But the Bahai Temple was the superlative. It took my breath away. I had seen the lotus flower structure in photographs, and fancied I had spotted it from the air just prior to landing, but as the driver rose to the brow of the hill and the temple came into view my hair stood on end. It stands triumphant and in silence, a monument to tranquility in the middle of India's capital. The gardens are simple and serene. It seems anyone who visits is immediately well behaved, they even queue quietly to get in. I enthused about it among my colleagues in the evening, most of them had never paid it any attention. I was surprised but when I thought about it, I realised I myself had never been to Buckingham Palace.

The sight I most wanted to see should really only be seen in the morning when the USA is finishing work and the US Shift at the BPOs knocks off. The Zaika Restaurant is legendary in the BPO world. This 24-hour *dhaba* has been granted a rare liquor licence and just as pubs around London's Smithfield Market are open at 6am so that market traders can have a drink before they go to bed, Zaika provides a similar service for the Gurgaon nightshift (and indeed day shift, evening shift and morning shift). Stories of drunken debauchery and bravado abound. If you want to stand up to your boss but need some Dutch Courage, this is the place to get it. If you don't want to insult your boss but can't prevent your own drunken sincerity, this is the place to avoid. This is where Gurgaon's call centre Bulls meet its Bears, where corporate moderation is flouted and excess is embraced. Permissive women are permitted, but it is very much the realm of men. Sarasvat had fuelled these legends for me and it was a shame he wasn't around to introduce it to me in person. He had given me directions on a menu he had swiped for me, in addition to its whereabouts, the menu read:

New Zaika Restaurant – Quality is our prime motto.

The only restaurant in town with 24 hours service.

OUTSIDE FOOD OR BEVERAGE STRICKLE NOT ALLOWED.

Rs.50 Surcharge on vomiting

TAKE CARE OF YOUR VALUABLE AT ALL ITEM.

Not a very good advert for the place in retrospect. How many restaurants charge you if the food is so bad it makes you sick? I explained to my colleagues that it would be due to drink that people vomited, but strangely they still weren't keen, I suppose they wanted their multi-cuisine menus and over priced, overblown, overtly safe Western haunts, and not a white face among them. I went alone.

I told my driver to wait and walked into a British Friday night in reverse. It felt like a busy curry house after the pub on a Friday night in England but one where everyone is Indian. I sat at a table thinking of the famous Goodness Gracious Me sketch where a group of British Indians go for an 'English'. This is exactly what was happening, groups of Indians behaving like their British counterparts - get drunk, have a curry, throw up. The waiters were being insulted, but in Hindi. The hottest thing on the menu was being eaten with relish, considered and then ejected into the toilets. Drunk Indians everywhere. I would love to have stayed until the next shift came in but I really didn't have the stomach - Rs.50 surcharges were being put on one too many bills.

****

The day to leave Pune finally arrived. I had arranged to bring a laptop from the office along with a wireless landline to Goa so that I could pretend to work remotely and in fact do very little. I was committed to this dishonourable plan, feeling very strongly that I needed to get paid for nothing as the final ignominious stage of giving up work. Leaving behind the IT department would be very easy to do. In fact, leaving Pune was very easy to do. While the city had its advantages and had provided us with some great experiences, it was a great deal less interesting than we had imagined. We felt quite sure that things would have been different if we'd been in Delhi or Mumbai or Calcutta or any of the larger cities, but Pune was a little too small in attitude for us and a move was long overdue. We would not miss 'cleaning woman', our unimaginative 'cleaner' who pushed a mop around our place occasionally, we would not miss the 8am sound of high-tech military aircraft loop-the-looping over the city, we would not miss the perilous motorcycle riders pulling out from all angles while we were trying to drive in a straight line, we would not miss hearing 'there's always Coca Cola' played electronically each time our neighbour reversed his car, we would not miss the bin-woman who rang the doorbell at 6am on each occasion that we had not put any rubbish out to enquire as to where it was (Julie tipped her against my wishes the day we left and told me gleefully once we'd set off) and we would not miss the neighbours' kids playing cricket until all hours in the stairwell of our block, shouting 'how was that?' or 'catch it' after every single under arm tennis ball bowl.

But leaving is leaving and it was not without its wrenches. Workmates would remain friends, indeed, we had already arranged for a Goa visit for many of them, but many of the neighbours would be missed, as would my gallantly abused tailor. We would miss the British Library, cricket matches at the Poona Club, the Non Veg Platter at 1000 Oaks, the mynahs and bulbuls (the most charming of birds), Kamal's cookery classes, the member's enclosure at the race course, the nasal cry of the newspaper wallah as he collected last week's tabloids, and most of all we would miss the best flat we had ever lived in, its whiteness and its unprecedented view of the sun setting through one window and the moon rising in another. And of course the exemplary Mr. Khan, our guide in all things mechanical and philosophical, hanging around his shabby garage, idly chatting with his customers and his antediluvian neighbour-mechanic about Triumphs, Morris Oxfords and decommissioned Unicef vehicles is likely to remain the most pleasant memory I will hold of my time in Pune. We would be driving the chariot he made for us all the way to Goa, a twelve-hour journey at the best of times, who was to say how long it would take in our little subcompact. Before I left, we took chai together in his preferred eatery, the Jubilee Rooms, where I assured him that he had made an indelible impression upon me as an exemplary businessman, and that while he had clearly fleeced me in nearly all transactions, he had done so in such a chivalrous and good mannered fashion that I felt evermore in his debt. He gave me a roof rack to help with the journey and despite the fact that we'd be in Goa, promised to buy the car back from me anytime I wanted as per our original deal. Somehow I knew I'd be seeing him again.

Laden with our sizeable accumulations our car looked pleasingly Indian. Thus encumbered it held the road much better too, I presume the six seat spaces are expected to be occupied at all times. We headed for Kohlapur where we intended to stop overnight at Shalini Palace, the decommissioned seat of the Maharajah of Maharashtra (try saying that with a mouthful of cake). Julie expressed a preference to drive and with consummate dexterity she guided us over mountain tracks with a seeming proliferation of suspension-destroying potholes for some six hours without noteworthy incident. Such was our confidence in our car that we defied the advice of all but Mr. Khan entirely, and pushed on through the hills to Kohlapur with the wind in our hair and huge smiles on our faces. As we stopped to buy hot *wadas* at a *dhaba* and watched tiny munias and sparrows pick over the left over rice on nearby tables I felt the anguish of recent months percolate through my body out into the hot shining day. Julie and I smiled at each other and began laughing. We didn't really know why, but the laughter went on and on. I told Julie about Puneet and his accidental philosophy, about the 'Eastern Sense' and how I intended to become more patient and we laughed a great deal more. We stopped by a river to watch a kingfisher and drink a Kingfisher, marvelling at a rickety transporter bridge like the one in Newport, and laughing at it. As we pulled in to Kohlapur and its chaos we laughed without restraint and I knew the Work and the city and the anger was gone, and that the fun was back.

Although we'd pre-booked the Maharajah Suite, it was occupied. Laughing, we took the Maharani Suite, a fine apartment resplendent with its own bar, balcony swing, a four-poster and a poster of the Swiss Alps (the ubiquitous Indian yearning). The bathroom suite was in faded pink and it might just have been the biggest bathroom I had ever seen. There was a bidet, the first one I had seen in India, perhaps the only place I can think of where a bidet should really be mandatory. The faded glory made us very happy and very at home, though the huge oil painting of the Maharani going into battle with gritted teeth on horseback wielding a knife was perhaps slightly unnerving. There was no toilet paper as (despite the fact that we'd booked ahead) they weren't expecting Western guests. The room service waiter handed me two rolls with a smile, left the room, put his tray down on the table outside and attempted to close the door exactly as I decided to walk out on to the balcony. I attempted to push the door open as he tried to push it closed. Thinking it was stuck he pulled it open and pushed it shut again, smashing my hand on the handle as I stumbled into the velvet curtain between he and I. The curtain got trapped in the door hinge and in an effort to free it, this small tough man waved his hands forcefully at the velvet, punching me first in the stomach and again in the side of the head. I howled in pain pulling the curtain aside and shouted "I'm trying to get out" as he again pushed the door closed by getting his whole body weight behind it. I tried the handle but he had it firm. I tried a surprise grab but he was resolute and I turned away, bruised and defeated, and resolved to stay in.

Given the tatty splendour of the room, this was no bad thing. Having enjoyed the waiter's last visit so much we decided to order dinner in the room, after all, we were in buoyant mood and ready for the fiasco of telephone ordering. On many occasions throughout the past year, I had revelled in the impossibility of room service ordering, in fact I had shot a number of short films of Julie's desperate attempts to be understood, films which, after our next experience, seemed like public information films on How To Order Room Service. We called for a menu that came immediately along with a request for both our height and eye colour. Bemused but unperturbed, we gave these details and perused the menu. When we were ready to order I handed over to Julie to do the honours.

"No, no chance. I drove all the way here, you have to do everything from now on."

I picked up the receiver.

"Hello, room service? I'd like to order dinner please. I'm sorry? No, dinner, I want to order dinner. OK? Good." I put my eyes up slightly at Julie. "OK, I'd like the chicken Kohlapuri, no, Kohlapuri, KOHLAPURI as in Kohlapur, the place where you live, yes Kohlapuri and one *malai kofte*, yes, *malai kofte, KOFTE* yes, and three plain *roti*, yes three, THREE YES, and one, YES THREE *ROTI*, and one plain naan, one naan, OK? Please repeat the order to me..." Rrrrrrrrr.

I replaced the receiver and turned to Julie, "Should be ok, but he didn't repeat the ord..."

Ding Dong.

I opened the door to find the room service waiter who had bashed me up beaming at me through broken teeth.

"Sir, please check order please sir." He handed me a piece of paper on which was scribbled:

'2 C Kohlapur, 2 Veg *kofte*, 2 R, 2 N'

I corrected the order to say 1 C Kohlapur, 1 *malai kofte*, 3 R, 1 N, with patience and a smile, shaking my fist only at his exiting back. I closed the door and walked back to lie on the four-poster bed. After ten minutes the phone rang.

"Hallo, Mister Mike?"

"Err yes…"

"Your order sir to confirm, one chicken Kohlapuri, one Veg *kofte*, three *roti* and one naan, is this correct?"

"Yes, more or less, that should be *malai kofte*, but no matter." I conceded.

"But you are not ordering mutter sir? You want *mutter*? Peas?"

For once I recognised the source of confusion (or should that be sauce of confusion?) "No, I said no *matter*, not no *mutter*, just the original order – chicken, *malai kofte, roti*, naan, ok?"

"OK sir, no problem, food is coming soon." He sounded confident and I again reclined on the bed to relax.

After half an hour a much-anticipated knock came on the door. I opened it to find two waiters with huge trays. They rushed into the room and I stood back to watch them unload a huge dinner on to the table, including two chicken Kohlapuri, two *malai kofte*, what looked like a *paneer mutter*, one dish which was certainly mutton, three *rotis* and one naan. I thanked the waiters enormously and Julie and I tucked in the napkins and gorged amid bouts of disbelieving laughter. Being polite and patient had become the new being rude and short tempered, and we were enjoying it.

We ordered wine with the meal on the strength that the hotel is now owned by the proprietor of Chateau Indage, and the wine list was all local. We took advice on one red and one white. They arrived. The red was cold and the white was warm. I called room service to point out the mistake and was told:

"Sorry sir, no white is cold, only Western guests are drinking."

I pointed out that the red should not be cold and smiled as I was told, "No sir, everyone wants all the wine cold only."

Fair enough mate, fair enough.

We set off early from Shalini Palace but it seemed the whole staff were there to send us off. As we checked out, the hotel manager walked with me to the car and asked me a direct question.

"Mister Mike, you are living in India no?"

I nodded.

"And you are working too, earning good money no?"

"The money is certainly adequate," I affirmed.
"Then why do you drive a car like this? I don't understand."
I appraised his well dressed, stiff countenance, groomed moustache, dusty shoes and beamed at him with sudden understanding.
"That's just it my good friend, we don't understand each other, we never have and we never will, and what is more, it is wonderful."
And with a tip of the hat to all, I slid my rich man's body into my poor man's vehicle and trundled off toward Goa feeling like I had it all.

After six bumpy, happy hours we conceded that we were lost. We had a map, but it was drawn up by a vicious liar as all Indian maps are. I asked directions at every junction, "Goa?", and they would point, mostly in the same direction. I didn't care. It felt like my efficient, worrying, angry self had been left back in Pune, I just felt like not caring and it was easy and I was enjoying it. After roughly ten hours driving on a six-hour journey we crossed the border into Goa. As we cut in towards the coast, heading through the familiar territory of Mandrem and Chapora, it felt like coming home. Our doubts about Goa, and indeed India, lay back on the road behind us, scattered and invisible among the red dust. The sun was setting as we slipped into the familiar and comforting homespun of the Casablanca Hotel. Gurmeet approached with a smile and outstretched hand.
"Glad to be back?"
"Yes, we're glad to be back."

Chapter Sixteen

Friends, friends, friends. Gurmeet, Anupam, Arnold and Isabel, Edwin, Santosh, and, of course, Peter. All still there; all welcoming us home. Now we just had to find a home, and remembering earlier times, it wasn't going to be that easy. We had decided to try another area - whilst Candolim was full of friends, it was also full of superannuated doddering Brits, and if we'd learned anything in the past twelve months it was avoid the whites and stick with the browns. We settled on two options: Reis Magos with its lightly lapping estuary or Arpora and its clandestine walled mansions. There were to be no more Oafs for us.

Unable to locate Jude, we found one Nobby, an overdressed man with ancient crash helmet and miner's boots who slimed easily into the role of Goan-style estate agent despite his apparel. He guided us round numerous Portuguese houses as per our remit, but they were all too big, gloomy or scary. We felt sure we'd eventually find one and were ready to put up with a few rats and snakes in return for a six-month slice of colonial history, but the notion of 'The Shining' kept slipping back into our minds again and again. With Gurmeet's help we found a nearly suitable setup in Reis Magos, a new build on the first floor with three water facing windows, but the concrete slab roof cooked it like an oven and the landlord and his family lived downstairs, they were nice, but not that nice. Our frustrations forced us close to compromise and we nearly took it, but as I gave in to laziness, Julie remained resolute and optimistic and set off to knock on doors in Arpora. Within two hours she had found the place.

Within a week we had retrieved our reposited paraphernalia from Casablanca's dusty store rooms, added them to our other mushrooming accoutrements and moved in to *Penedos Da Soudade*, a 350-year-old Portuguese mansion. The place was secretly slotted down a track leading to the river and almost completely covered by the forest canopy of palms. A real village retreat with hardly any passers by save for buffalo farmers, fishermen and lost tourists. There were neighbours on one side only, a family who had once been servants to the mansion owners and who had lived there ever since. This really was a property of some magnitude, five bedrooms, all en-suite, two gargantuan halls with eight metre high ceilings (yes really eight metres), a vast dining area and a garden compound of some thousand or more square feet, possibly even double that (a surveyor I am not). Our new landlord and proprietor one Ramiro Pinho, an undersized descendant of the Iberian subjugators of the region, gave us a full history of the house inherited by his great grandmother, along with a copy of a book of his reminiscences. He was delighted that we intended to spend our time writing and painting, being a keen advocate of both himself. Like so many Indians we had met, he placed honesty at the forefront of his list of required attributes, reassuring us time and time again that we could trust him. I have no idea what impression most foreigners have of Indians, but it must be generally bad, why else would everyone feel the need to persistently defend themselves against accusations of fraudulent knavery before any are even uttered? While giving us a tour of the acreage, he was at pains to identify a 'scruffy' area in the grounds that he recommended we maintain unkempt as a holding area for snakes. I disclosed to him that we had witnessed a medium sized python at close quarters only days before and he assured me that a comparable occurrence was entirely likely within his grounds were we not to heed his advice.

He had erected a pleasing summerhouse towards the rear of the estate, which proved ideal for Julie's painting activities. I counted thirty five coconut palms, three established mango trees, two jackfruit bearers and over twenty other food bearing trees and plants within the grounds and our custodian assured me that he would be along at regular intervals with his staff for 'plucking', it seemed he has a sister with a particular fondness for mangoes which he imparted was a preference that it was 'more than his happiness was worth' to ignore.

The kitchen was open to the elements with only some painted ironwork to guard against sentient (warm and cold blooded) raiders. Ramiro informed me that he had to seal one alternative exit to the garden after repeated incidents with a medium sized monitor lizard, the self same having developed a predilection for a former tenant's rations. But after a month we had had no such incident other than the inevitable mosquito barrage, largely handled by the adequate battalion of gecko lizards. We did however experience an ongoing plethora of ants attaching themselves to everything edible with a remarkable doggedness. On one occasion I drowned some 300 or more of the blighters in a frying pan at which they had been drinking remaining oil, arranged in a perfect circle, surrounding their quarry like Sioux Indians or too many wolves.

It felt like we had settled in overnight. Perhaps it was our mood, happy again, happy to be back. The extent to which it felt like home, and so soon after we had expressed serious doubt about staying in India for even a day longer, was slightly shocking. We began to wonder if we'd ever go back.

The height and overall bulk of the building, and the steeply sloped roof of hard fired Mangalore tiles meant the September/October stickiness that is Goa's humid post-monsoon was kept almost completely at bay, but the torrents of the post-monsoon showers were certainly not. And although they came on an almost consecutive basis each night for a week in September, we always seemed to be taken by surprise. First the electricity would go out, struck by lightning. Next the thunder would scare us to death as we scratched around hopelessly for candles in the dark, in our underwear. If we were lucky we would have enough light to find the buckets before the heavens opened and the carefully placed roof tiles displayed their gaps like proud faucets. We would then rush about in soaking underwear desperately emptying buckets and slipping barefoot on the by now soaked terracotta floor. Given the size of the building, we overlooked our displeasure at having servants in the house and employed Siddarth from the Casablanca Hotel who would come for an hour or so each day apart from Sunday when he would play football. He showed no more initiative than any other cleaner we'd had, but he came most days, did whatever we asked and increased the building's sparse population by fifty percent.

The sheer corpulence of the place allowed us to breathe to a degree unprecedented in both of us. We even accepted the countless inescapable livers-in that we had to accommodate, the imperious jungle crows with their incessant squawking were the most brazen, barking orders at each other and taking the kitchen by storm rather than stealth. On one occasion they pilfered two full chicken breasts (in a plastic bag) and eighteen eggs over a three-day sustained assault. The squirrels could never manage such military pluck, but settled surreptitiously for the vegetables.

And we were visited by the most fantastic array of birds one has ever seen. While most of them were way up in the canopy, many would visit us at ground level for an inquisitive poke about, most notably we made something of a pet of a male paradise flycatcher with its impossible rainbow of tufts, curls and tails. It transpired that our patch was in fact a bone fide bird watching Mecca and by early December we would be occasionally joined, on the other side of the garden fence, by binocular-ed bird buffs clocking the feathered acrobats on our mango tree as we lounged on the spacious verandah drinking our morning tea. We saw everything, from parakeets and pittas to barbets and bee-eaters. On one occasion we had to duck and cover as an Indian eagle owl sought screeching refuge in our porch rafters from a mob of at least ten house crows chasing it with serious intent. That genuinely scared me, an Indian eagle owl is a big old bird.

We had to do something about the mosquitos, they were feeding freely on us and getting so large that, to steal a quote, you could 'freeze dry them and use them as deckchairs'. Getting feet mixed up with metres, Julie came home from the tailor with a mosquito net the size of a small house. Fifty metres of cloth goes a long way, and having twenty or so dragging on the floor around the bed proved pretty effective in stopping the bastards getting in.

So we selected a studio each and spent our days in semi-seclusion, busily working at the things we had set out to achieve. And it felt very good. We built up a routine, something we thought we were trying to avoid, but it made things easier, taking away the need for constant decisions. Julie began to curate shows at a local art gallery, including her own paintings, and they were well received and started to sell. She was finally able to locate a small group of local artists and start working with them, bringing Indian themes into her own work and exhibiting with them. She churned out paintings and I churned out words. Things we working out. Things were finally working.

But not everything worked, this was still India after all. The car began to play games. Indian engineering being what it is, transporting a vehicle from one place to another does not mean that said vehicle will subsequently work. The salt air played havoc. The float pin in the carburettor was getting far too sweaty and not taking in petrol at all and the exhaust was sounding like an Eton schoolboy's raspberry each time I accelerated. In my efforts to find a mechanic I went through too may chumps to name, nobody in wealthy Goa was interested in tinkering about with an *Old Fiat* as they dared to call my shining *desi* vehicle. In truth no one had a clue how such a master's chariot worked. Eventually I became availed of the services of one Dilip, a bodge-master extraordinaire who finally transformed the vehicle through the laborious method of trial and error. While he was dismantling the engine piece by piece I hung around with his kids playing *tablas* and constructing impossible toys from home made Meccanno and pieces of string. The 'superbike' they were building from a child's cycle and the fairings from a scooter made me want to kiss their father's influence.

Our efforts at finding new European friends didn't really work either. We did meet a few people who we started to knock about with, but their inability to talk about anything other than tattoos, drugs and dogs, and our total disinterest in talking about the same meant our efforts were never really going to come to anything. But Julie's new-found art crowd became good friends of ours and our social horizons began to broaden a great deal with trips to Mumbai and Delhi in prospect and introduction to a host of less touristy, 'local' bars and restaurants within Goa, ones that we had never stumbled upon ourselves and probably never would.

And the TV didn't work. This may come as no surprise as the TV in India rarely works, but the problem was not the cable operator, but the cable operator's operator. He just simply didn't work. His mother must have predicted his indolent demeanor when she christened him Eulazio. Julie and I immediately christened him You Lazy Oaf but after the fourth week of failing to turn up and re-attach the cable to our TV since it had been detached by a passing lorry, we re-Christened him You Lazy Twat.

But most fantastically of all, I didn't work. Well, I worked on my writing, but I didn't work on my Work. I wasn't sure if this was because nothing had changed with respect to the clients and therefore there really wasn't any work to do, or whether it was due to my own tenacity in adherence to my work-but-don't-work plan, because I had really stuck to it. I had shown almost no interest in anything the management had asked from me. I had dealt thoroughly but succinctly with requests from my colleagues and ignored altogether requests to do with administration, which, as I suspected, eventually just went away if ignored. I largely avoided the fortnightly 'meetings' on the telephone by either just not picking up the phone or joining the conference call and pretending to get cut off. They tried to get me to the office on numerous occasions but I realised if I didn't respond to requests immediately, their forward planning was so poor that within a day or so the opportunity to make it for meetings, etc would pass by, after which I would send my apologies, blame Indian technology and ask them to send me the notes relating to the event - sometimes they even sent them, but I never read them. All the while I was being paid a salary. A meagre salary albeit, but a salary nonetheless, so eventually guilt got the better of me and I resolved to 'do some work'. I instigated a training session, stating that 'since there is little for me to do on the project, the least I can do is impart my experience to others'. I had assumed that this sanctimonious crap would assuage both my own guilt and any accusations of my lassitude that might be landing upon my boss. I copied all the *big swinging dicks* on the email and got an almost instantaneous response from the CEO saying something like 'we are not committing any further revenue to this project until we see substantial investment from the client.' Fair enough I thought. In fact *nice one centurion* I thought.

As time progressed and Christmas arrived I felt that I really had kicked the need to work, but the money was bugging me. It was all very well saying I had quit, but the money said I hadn't. I had cut down but I hadn't quit altogether. But to jack it in would be foolish surely? This is money we're talking about. I resolved to do no further work and get paid until they decided to stop. I called Avinash and told him the same, maybe I dressed it up a little better, but I gave him the gist. His reply was the green light - just hang on, if nothing happens so be it, if it does, take your decision then, we owe you that much. Sounded fair enough to me.

****

Christmas came and went peacefully. We avoided the neighbourly fuss of the previous year, electing instead to stay home, cook Indian food, get a little drunk and look at our six foot Christmas tree being dwarfed by our enormous home. For New Year, Manohar arrived. He was supposed to come along with Sarasvat, Shanti and Vikas from the office, but the others declined at the last minute, so he came instead with Anita, his twenty three year old girlfriend. Her father was apparently under the impression that she had gone to stay with a girlfriend's family, and his daily calls to check all was ok reminded me of being thirteen again. For New Year's Eve we had decided to go to Anupam's new place of work. During the tourist off season he'd been tempted away from his employer of nine years to a new 'venue' which had opened up directly on the beach, and New Year promised to be a rip-roaring night out of music and dance with the promise of International DJs no less. As we arrived the place was silent and in darkness. This was unexpected, by 11pm one would expect the majority of guests to have arrived, even in Goa. Instead there was emptiness and tales of a catalogue of disasters beginning with a failed generator and culminating in a police closedown. We listened to the stories of disaster, feigned excitement at a dull display of seasonal pyrotechnics and left. There was only one thing for it, Disco Valley.

The beach parties in Vagator need no introduction from me, at thirty-five I really don't have the stamina for raving, but a desperate situation called for desperate measures. Manohar and Anita said they were 'interested' but didn't know what to expect. We located Peter amid the throng of revellers on the main road in Candolim who were wondering what to do now that the fireworks were over, and we set off. The roads were mayhem, everyone it seemed was looking for somewhere to go with no idea of exactly where. Peter tried numerous short cuts to get through the queues but to no avail other than to make us all royally travel sick. After and hour and a half in his tiny van we arrived, precisely it seemed, as did everyone else. As we stood in the queue the music was deafening and all around us people were throwing up from carsickness. We joined them. By this I mean we all threw up, all four of us, not a good start. After a short queue, dodging little piles of puke, we got in. It was 2am and the party was on. Pounding techno, fifteen-minute tracks, boom, boom, boom, boom and the crowd up to its eyeballs. We'd had about four pints and most of that was now in the car park mixed into our digested dinner. We bought drinks and wandered about saying

'Err..." and not much else. After half an hour we'd simply had enough. Manohar and Anita were 100% uninterested and try though we did, Julie and I just weren't bothered either. At 2.30 we left. Amateur night. I've said it before and I'll say it again.

In January, the squirrels moved in. There are two types of squirrel in India, the 'five striped palm squirrel' and the 'three striped palm squirrel'. They are called palm squirrels because they often live in palm trees. They are small and cute and for the most part very pleasant. But in January everything changes. As mating season approaches they start making nests and start making noises. From 4 to 6am and 4 to 6pm from January through to March they squeak in a very loud, high-pitched, unrelenting way. Earplugs are sufficient when the squirrels are outside, but our squirrels were inside. Five pairs of squeaking, squawking, beeping, shouting, never-shutting-up balls of frivolous fluff made their nests in our rafters, rafters that were *inside* the rooms. At first we found this most endearing, laughing as they stole cloths and toilet roll for their nests, giving them names and enjoying their brazenness. We even played a little game, jumping up to startle them as they began a ten metre journey round the electric trunking, holding on with one arm and one leg, unable to turn and obliged to speed on to the end of the track despite the danger. But the mornings became too much. There was nothing we could do, we didn't have the heart to poison them, and hitting them with sticks just made them squeak louder. One morning I awoke with a bad temper, rose and started shouting back. They stopped. It was the breakthrough we needed. They would shut up if it seemed a rival was near. It seemed absurd to me, but it seemed they would be quiet if threatened by an interloper. But after a couple of days they sussed me out, I just didn't have a convincing enough voice. Then Julie had an idea,

"I've seen this toy in the market..."

She returned from Mapusa with the most wonderful example of bad craftsmanship imaginable. It was a pathetically formed six-inch plastic red goat with partially detached legs, a large rubber bladder, a pipe in its arse and a huge yellow feather on its head. It worked on the principle of the 'jumpy spider', where air is pumped into its legs via a rubber pipette on the end of a flexible pipe, the legs inflate and the spider jumps. This creature's mechanism differed slightly, here the bladder inflated and pushed against the otherwise static legs forcing a sort of shuffling motion that presumably passed for leg movement among the unconcerned. But here was the thing, as the air was pumped into the bladder, a high-pitched squeak emitted, a squeak very much akin to that of our furry friends. We took it into the garden for testing and Julie held the red goat aloft and waited expectantly for a squirrel to oblige, her eyes darting this way and that in serious anticipation. Soon enough a stripy rodent set forth on a spree of noise. Julie pumped the pipette for all she was worth, holding the plastic goat by the legs so that the former part of its body jerked forward and back, causing the feather to sway erratically and the high-pitched note to sing out in short steady bursts. The squirrel stopped. And it didn't start again for some minutes, as soon as it did, Julie was ready with the strident goat, it stopped again. This time it stopped for good. Total success. As January progressed the toy became invaluable, and mornings went something like this:

"Squeak, squeak, squeak, squeak, squeak…"

"Julie?"

"Mmmmnnnn?"

"It's 4 O'clock. It's your turn…"

"Huh? Oh no, can you go? I didn't sleep too well."

"Oh, ok, but you owe me…"

Rustle, rustle, scratch, yawn, pad, pad… Squeak, squeak, squeak, squeak, squeak…pad, pad, yawn, scratch, rustle, rustle, ZZzzzzzzzzzzzz.

Chapter Seventeen

February promised to be full. After nearly five months of working-but-not-working we had plenty to show. My book was taking shape and Julie's paintings were packing the gallery, but we needed a break, and it came in the shape of friends from home.

Max and Mandy came back. They'd tried to keep away but India had got them. They brought Malcomb and Sarah with them, two debutantes with a lust for the spirit of the East. They arrived on the charter flight and it took them a couple of days to get their bodies back in shape. During this time I talked at them at one hundred miles per hour without a break. When they began to visibly flag I offered an apology for my 'Me Me Me' deluge as having been holed up in the house for so long in front of a computer with no-one to talk to but myself, and occasionally Julie. I needed a real audience and finally I'd got one. I also needed a really good laugh and told them so, they felt sure they could oblige. But I also need to finish my chapter so the next day I sent them out to the beach while I hastily scribbled down a few last words. On their return I got the big laugh I was looking for, four dumb Brits stood in front of me burned red from head to toe. Over the next few days we went to the beach each day, somewhere I really hadn't been for months apart from the occasional evening walk or pint. In fact I hadn't noticed how much stronger the sun had become since Christmas and took a fair sanguine myself. I took them first to Santa Monica Beach Shack in Baga principally because they made great chips. Next day we went off to Mandrem for a day of classy contrast at La Plage, lounging in middle class French comfort, eating chicken livers and herby scrambled eggs. After the this we spent a day in the Calangute-Candolim wilderness, that part of the beach too far from any beach road for tourists to walk, and subsequently empty, it must be at least fifty metres of expanse, at least.

Malcomb determined to have his winter hair cut back to facilitate all over brownness, and being an advocate of the 'loads on the floor' crew cut myself, I offered the use of my electric hair clippers. Sarah and I took turns in trying to shear and prune his sturdy locks, but of course electricity supply is sporadic at best and hallucinatory at worst, and on this day the power was only strong enough to half cut, half tear the hair for a few minutes after which it dropped to levels insufficient to even remotely finish the job. With half a haircut we headed out to enlist a local barber. Said coiffeur engaged his tonsorial equipment with some speed but as the haircut was nearing completion the power failed. The industrious barber shrugged his shoulders, took up the scissors and finished the job in darkness. The morning of course yielded many lopsided laughs. Two weeks later, strangely undeterred, Malcomb visited another barber near our former Candolim residence. Were I in accompaniment, I would have advised him to avoid this *salon de beaute pas* at all costs, having had a previous run-in with the destroyer therein wherein he gave me a number three crew cut on top and then inexplicably started shaving off the sides with no comb attached. Malcomb's experience was predictably not enjoyable. The clippers snapped at the half haircut juncture and despite Julie attacking the man with her fists, the clippers could not be fixed. After bilious expletives from a furious Julie, the bewildered Malcomb was whisked away to our saviour barber who, upon seeing Malcomb with half a haircut for the second time in as many weeks, batted no eyelids and asked for no explanations, simply doing what he could with his clippers until the power went off and he was again forced to finish off badly with scissors. Malcomb asked me later in a sad, boyish voice why Indians found it necessary to always give two haircuts instead of the conventional one. I simply patted his bemused, wonky hair in consolation.

   We took a trip up to Bombay to stay at Shelley's Hotel where I had been so inspired at one time that I'd proposed marriage to Julie. Taking the overnight second-class sleeper train served up our newcomers a taste of Indian travel and instilled us all with a haven't-slept tetchiness that seems to precede any visit to India's tinsel town. Arriving at 8am doesn't help either. As Colaba awoke and gave itself a shot in the arm (in many cases quite literally), we wandered around in a daze of befuddlement. Anywhere else in the world we would have been ripped off - mugged, pick-pocketed, swindled or at the very least hassled, but Bombay is occupied with itself, and perhaps a strain of honesty long forgotten elsewhere. We waved away the huge balloon vendors (huge balloons not vendors) as they approached with a stoic smile, slapping the enormous balloon held out in front of them, "Balloon Sir?" We cold-shouldered the shoe shiners and ear cleaners with their half-hearted approaches, and we showed the already zealous marigold sellers the palms of our hands. At 9am I called Shelley's to plead and they let us in. We slept until 3pm, rising in time for a late *pav bhaji* lunch at Badshah's opposite Crawford market before heading into the market to find some readymade homespun or *khadi*. I had a notion that we could have some fun if we dressed up in *khadi* outfits and Nehru hats. I had seen countless Westerners in tie-dye and pyjamas, not to mention the sari-bindi tourists, but I'd never seen white people in the plain white garb of incalculable Indians countrywide. The problem of course is that such garments are made by local tailors, the cloth is available, but the outfits are not. Having been on every goose chase possible, I could see interest in my idea waning, so dumped the others at a *falooda* bar and followed the most adamant of our hassle-guides to a shop he irrefutably insisted had Nehru hats at least. He was right and I was able to secure a number of the same at the greatly inflated price of 100 rupees each (I was sure they would normally cost about Rs.20, but how many times is this tailor going to experience a white man being led into his Lilliputian shop with an express order for the cheapest headgear imaginable? One has to pay for context). We donned the hats and walked into the street. The reaction was immediately astonishing, we were accosted from all angles, first I worried we might have offended in some way, perhaps breached some unwritten religious taboo or just been seen to be taking the piss, but I was wholly wrong. We drew only laughter of the inclusive kind. As we walked people stopped us to shake our hands, people who did not speak any English and had nothing to gain from us patted us on the back and smiled as we passed. One man wearing an almost identical white hat stopped me in my tracks and held my hand firmly, smiling,

"The hat you wear was worn by our Great Jawaharlal Nehru, the first Indian Prime Minister, to see a foreign man wear this makes me proud, for this I thank you."

We decided to put the hats away, after all, we were only trying to have a bit of a laugh.

Our artist friend, Dr Kerkar was putting on an installation piece on

Chowpatty Beach similar to one Julie had helped him complete at Miramar to coincide with the International Film Festival of India back in November. The spectacle comprised a number of enormous holes in the sand lit by bare bulbs, myriad clay nipples of really some magnitude, an enormous bamboo gallows strung with saris, some twenty thousand green lipped muscle shells and a most fantastic arrangement of interior lit vulva-like constructions which I made the terrible mistake of likening to surfboards. The locals seemed unsure quite what to make of things, but the art crowd was nothing less than delighted, and Subodh got himself a Times of India front page photo. In celebration we headed off to the legendary Leopold's for a night of truly stultifying AOR music in the earsplitting upstairs bar. It got so bad, that in the spirit of doing so many things he'd never done before, Malcomb asked the DJ if they had anything by Chicago. The very next song began "If you leave me now, you take away…" but we left before hearing anything more.

On the night train back to Goa our guests reflected on their Indian experiences so far, shaking their heads with astonished smiles. I listened while they enthused about Cuffe Parade and its majestic Art Deco, about the racecourse, the *dhobi ghats*, the smell of jasmine and incense and the sea, bombay duck at Mahesh Lunch Home and king crabs at Trishna, the terrible Insomnia nightclub at the Taj Mahal Hotel, the *heedra* who attacked us in Chor Bazaar and after hours drinking at Gokul. I listened as my own India opened up to me again through their eyes.

I stopped them and offered a quote I'd written down from Rushdie.

'A kind of India happens everywhere; everywhere is terrible and wonder-filled and overwhelming if you open your senses to the actual's pulsating beat.'

Malcomb nodded, "Yeah, that's it mate, that's it."

****

Back in Goa we got back to the task of having a good old British holiday, just enjoying the sun and the time off. We gave the other British a few lessons in the art and etiquette of consumption (not TB, the Brits don't have that, that's mostly the Indians) and how to behave when and after indulging in it. One notable occasion on a 'drinking boat', an hour long entertainment cruise up the Mandovi from Panjim, saw vigorous dancing actions to the excellent 'Birdie Song', which has by all accounts become unfashionable among British tourists these days. I am certain that we have again made this song fashionable, particularly the bit where you all join arms raised aloft and go round in a circle, the bit where it goes 'der ner ner ner, ner ner, der ner ner...' everyone else had forgotten that bit.

Sitting in the summerhouse eating breakfast and writing postcards, Malcomb was appraised angrily by a noise from the rafters, a Cobra. A seven foot long Cobra. He called the rest of us. We agreed, definitely a Cobra, just about the most aggressive and deadly snake on the block. As we stood at a good distance, worrying that it might decide to go into the house, a fully grown male langur monkey appeared on the garden wall in front of us. We traded surprised glances as it began to gesticulate first with wild head bobbing, then in a menacing bouncing-up-and-down manner before letting out a few sharp screeches and running the length of the wall on its hind legs to dive headlong into a cluster of bamboo and shake it about a good deal. Before we really had time to digest the garden menagerie further, the snake piped up with some strange coughing noises and determined to turn its enormous self around, the better to see the whites of our eyes. As one, we stepped a little further to the right. The monkey ceased its rustling, leaped back up and over the wall and advanced towards us, stopped only momentarily to semi-demolish a small banana tree. Snake to the left, monkey to the right. Silently picking up stones and forming a tight group formation, we edged quickly to the garden path, turned and ran, reaching the kitchen and bolting the door in seconds.

"Err...perhaps a day at the beach?" I offered as everyone was running out of the front door grabbing towels and sun lotion.

After three weeks, our friends had to leave. The day before their departure Malcomb and I were strolling along the river at the end of the lane and caught sight of a herdsman as he walked barefoot into the river, dutifully pursued by his fifteen lolloping buffalo. He crossed the river without even acknowledging that he had got wet, just as if he were crossing the road. The filthy buffalo were much more enthusiastic, bellowing brays as they marched through the water and emerged clean. Simultaneously a flock of rose ringed parakeets descended squabbling upon the palms. The sky was cloudless and the sun shone directly above. Malcomb confided in me that he didn't want to leave the next day.

Julie and I didn't really want to leave either, but time was pressing on. Our tickets were booked for March 21st, some three weeks away, and we were flying from Delhi having decided to take a trip to the north for a quick peek at what we could expect the next time we visited, and a way to occupy our final days to somehow soften the shock of leaving. Our itinerary had been booked through Marlon our trustworthy travel agent. Despite not trusting any other Indian travel agent, we'd come to trust him after he came up with a bus ticket for Manohar at New Year, if he could do that, he could do anything. And we'd also learned that doing anything for yourself in India is futile - don't feel guilty about it, just let someone else do it, they're the experts. The itinerary would include a flight to Delhi where we would stay with Shanti's family for a few days, an offer made to us some time back in Pune, and one we felt we must take up, after which we would take a train to Chandigarh, then Shimla and finally Agra before returning to Delhi to recover and fly home. 'Fly home'. The phrase had a ring of terror about it, an electric shock about it, a jabbing sense of fear. We would be going home at last. We didn't know if we wanted to. What would we do when we got back? Where would we live? How would we live?

Suddenly we had only a week to reflect, a scary, anxious and strangely melancholy week. What had we learned from India? What had we achieved? Should we be leaving at all? Why didn't we stay on? We'd fallen in love with Mother India, why desert her? These questions tugged away at our shirtsleeves like begging children, niggling us with doubt. We tried to exorcise our demons with lists, lists full of reasons for and against. We wrote them in, scratched them out and then wrote them in again. But it was futile, this wasn't a time for reason, it was a time for gut feeling and facing the truth. We had not fallen in love with India at all, we had fallen in love with *our* India, our safe, English speaking, middle class, beer drinking, unBelieving, version of this land of paradox. We somehow knew that if we stayed on, we'd get bored and have tantrums. We would get cross when the TV didn't work and our parcels were lost in the post. We'd get angry when people stared at us in the street, and when they persisted in their begging. And we would never find a way to stop the mosquitos. India was not really a place for us to *live*, we are Brits, we belong in Britain.

So as our leaving approached we thought instead of home, of missing our families and friends, of solitary walks along the Thames, the sound of a blackbird, the smell of fresh tarmac or creosote, of Tower Bridge, fish and chips, chicken tikka masala. And we focused on the positive things we had achieved. I looked at my list:

"…take up Yoga, play darts, tinker about with old cars, play guitar, get drunk, sunbathe, learn a language, keep a garden, look for some way to help out in the local community, muck about, improve our photography, watch hours of cricket"

…well, we had certainly developed a suntan. But we had changed our outlook on the world. We had run away from the world of Work, and although it caught up with us once or twice, we had finally turned our backs on it. I felt certain that back in England I would not succumb to Pacification by Salary again. I looked at the nearly finished book in front of me and thought, "I'm a writer now, at last, I have written something", and I felt good. I felt I would do something with this writing from now on, and that I'd no longer waste so many hours, days, years on chasing the Work. I'd have to do something to support myself, but it would no longer be *that*, it would not be *that institution* any more. To confirm myself as ex-addict I called Avinash and told him I no longer wanted to work for the company, despite murmurings about me continuing in some capacity from the UK. And once again he accepted my words and wished me all the best.

"Avinash, this time I mean it you know, I'm really not coming back."

"I know Mick, I know, but just keep the phone on until you get back OK? It won't work in the UK anyway, but I'd at least like to call you once again before you leave."

"Fine, but don't get any ideas, this time is for good ok? All the best…"

Having kept ourselves to ourselves since coming back to Goa, we were able to escape with the minimum of fuss. We had grown tired of goodbyes, they had ceased to mean anything. There was one exception,

Mr. Khan. We had sold our fridge to Ramiro and the TV went to Anupam, but no-one was interested in the car. I called Mr. Khan and explained the predicament.

"Mike, we had a deal no? I will buy the car back from you as stated. I am a man of my word."

"But I can't drive it up to Pune, how will you pick it up?"

"I will come myself to Goa and drive back, it will be fun."

"Well, if you're sure…"

Mr. and Mrs. Khan arrived on the overnight bus from Pune, picked up the car, paid us and drove back again. I asked him if he really wanted the car back or of he just felt some sense of duty over our agreement.

"Mike, it gives me satisfaction to do such things, the motive is immaterial. You have been fair with me and I with you, I hope we can deal again in future."

As we waved him off I sincerely hoped so too.

The next day we packed up our memories, renounced the remainder of our accumulated possessions to the nuns at the orphanage and set off to Delhi. Peter wasn't too sad to see us go, he shook our hands and said, "you'll be back."

****

We arrived in Delhi and dragged our badly beaten coffin sized trunks through to meet our waiting driver. I knew that Mr. Khara had gone to the trouble to print out a placard with our names on it but it was nowhere to be seen. I went up on tiptoe and wobbled my head around a great deal to motion to the hundred or more drivers holding placards that I was looking for my name. As I walked along the line, a driver leant across the railing pushing a placard in my direction with the words 'Ravi Shankar' written on in felt tip pen. I laughed out loud and commented:

"Ravi Shankar? No mate, you're looking for a skinny octogenarian Indian with a big funny shaped guitar."

A big laugh went up among the drivers, surely they can't have understood what I'd said? As I progressed along the line I found it impossible to keep my composure when driver after driver called "Ravi Shankar?" to me with hooting laughter, I realised it wasn't my joke at all but the comedy of the initial enquiry - asking a white man if he was Ravi Shankar. After walking along the railing twice in absolute hysterics, I finally spotted my man with our names on a placard upside down and managed to curb my cachinnations for long enough to nod and give him the thumbs up so he could come to rescue us from the "Ravi Shankar" screaming mob. It took Julie and I a good twenty minutes to stop laughing with many reprises and lots of 'only in India' sighs.

The Khara house was nestled amid a maze of walled gardens. An eight bedroom castle that should have been Mock Tudor, but wasn't. This all-white alcazar bore more than a passing resemblance to its spiritual brothers and sisters in the English County of Essex, and, Punjabi through and through, it sat behind a beautiful hand-clipped lawn on which its most genial owners sat waiting to greet us. We did the hand-shaking-while-putting-the-other-hand-behind-the-back-of-the-person-you're-greeting thing that Indians do so well and in which I often get all mixed up, and I wondered if I was expected to do the shoe-touching thing which always alarms me whenever I see it.

The Khara family are true Delhi Punjabis, losing everything at Partition and gaining it all (and more) back after relocation through a mixture of dogged business acumen, determination and hard, hard work. And Mr. K was certainly no slouch, owning a great swathe of prime Delhi building land as he now does, not to mention a series of very serious money making businesses. They also have the Punjabi hospitality that is unsurpassed the world over. They had provided us with an en-suite room, room service, a car and driver and a calendar of social events to attend every night of our stay, for which, we were informed through knitted brow, we would contribute nothing but our presence. We were taken to dinner at the golf club, a garden party thrown by someone apparently at the top of the Delhi glitterati pile (we had no idea who they were but the newspaper photographers seemed very keen to know who we were), lunch at the Habitat Centre and on the last couple of nights we had been invited to the wedding of a family friend, at which Shanti would join us. This is not the sort of itinerary one expects when dropping in on almost total strangers, I have genuinely never been made to feel more at home by people I don't know.

After a couple a days looking around Delhi and a day being taken around the commercial art galleries by Shanti's art-restorer sister in law, Padma, we finally got the chance to take a driver out to a place we had read about and were impatient to see, namely the Sulabh International Toilet Museum. The driver was unsure as to the location but after some three hours of gridlock and asking people in the street, we arrived. The driver's eyes nearly popped out when he read the sign saying Museum of Toilets. The place is a marvel, the tireless work done by the organisation to lift the dignity of the human refuse collecting *dalits* is outstanding. Fee paying public toilets (1 rupee) fund a school for 'untouchable' kids where they are taught academic and practical skills to lift them out of their 'undignified' profession. We were taken round the school and introduced to the children, given a fascinating tour of the laboratories to see experiments in the biodiversity of human waste, and finally cooked piping hot *rotis* on a methane flame, powered by human excrement. The International Museum of Toilets itself houses all manner of lavatorial porcelain and toilet related humour including a properly rib tickling scroll of historically famous words on the subject of excreta. While we were having the tour, our driver came in to check that his eyes had not deceived him and that we had genuinely come to a museum of toilets. He left in bemused giggles. When we finally returned to the house, the driver rushed off to the servants quarters to tell them all where we'd been. The servant's laughter could be heard through the house until the cordial and avuncular Mr. K boomed his displeasure from atop the sweeping staircase.

After three days lording it up with the Kharas, we set off on our grueling railway run. At the station in Delhi, the recorded train announcements start and finish with a flourishing 'Der Ner' from an electric piano that might have been used by a 1970's magician, making the trains seem rather special, perhaps as though they had appeared by magic: "The next train to arrive at platform five is train 2058 from Chandigarh, Der Ner." Such finesse. The modernist in me was keen to see Chandigarh, the showy new capital of Indian Punjab that had been purpose built after partition. After a false start where Albert Mayer and Matthew Lowicki's leaf pattern for the layout of the city was abandoned due to Lowicki's sudden death, Le Corbusier was awarded the task of taming India with modernism. The moment we arrived we realised he had succeeded. As this was our first time in the Punjab, it was impossible to say whether the discipline we were observing on the roads was a result of the regimented grid network, or simply down to a different standpoint among people who were noticeably beefier, more boisterous and better turned out than their southern countrymen. There was something of the European in Chandigarh, something decidedly homely, and not just its Welwyn Garden City-ness. From the train we had seen familiar agriculture (wheat), pylons and the odd grassy dale, pleasing sights readying us for home, and in town we met a number of young Punjabi men from Birmingham with a keen interest in beer and a strong dislike for Indian food; as they followed us down the street in a souped up Contessa, tooting the horn and shouting from the windows as heavy beats boomed from the stereo we began to feel most at home. And the Punjabis love their shopping too, it could have been Saturday night in Central Milton Keynes - a stroll around sector 17, taking in M&S, Next, Diesel, Levi's and Bennetton before loading up on popcorn and heading into the cinema (OK, Milton Keynes' cinema wasn't designed by Maxwell Fry). The city's 'sectors' (in theory each containing its own maintenance organisations, food provisions, parks and schools for its 5,000-20,000 inhabitants) are flanked by larger gardens and a manmade lake spread over three square kilometres to the north of the city.

Hugging the lake is the world famous Rock Garden, an organic hotchpotch concrete wonderland of canyons and caves, rivers and waterfalls, strewn with peculiar human and animal forms sculpted from broken ceramics, glass, plastic bangles, wire and other discarded building materials by Nek Chand, an ex road inspector for the city and public works department, who, fearing ridicule, kept the whole thing secret for many years. Finally it was discovered by the authorities and to his surprise and distinction, opened to the public as a respected national monument. But the Rock Garden is not my favourite place. Having entered with just over an hour before our train for Shimla was due to depart, and having left the driver with strict instructions to keep the engine running, we sped through the maze of visual weirdness with a cursory interest and an eye firmly on the clock. After half an hour in the labyrinth we were less than half way through, lost and wondering if we'd be able to find a way out, beginning to get a little desperate perhaps, the bizarre sculptures seeming to mock us at every turn. Unable to see a way out among The Way Out, and getting similarly confused looks from the other excursionists, I clambered up to a rocky vantage point and located the exit. We surmounted two flinty walls before being apprehended by an affronted attendant and slung out on our ear…out of the exit and into our waiting car. Appallingly, necessary behaviour.

The train took us to Kalka where we made a connection with the toy train that would haul us very slowly up to Shimla, the summer capital of the Raj. Of all the hill stations and their inherent Britishness, Shimla is the most reminiscent, the spires of Christchurch reach up into the Himalayan sky saying 'Oxford, Oxford, Oxford'. It is said that at the heyday of the Empire, one third of the world was governed from Shimla (or Simla as it was known then).

On the toy train, legroom was none. As on all the trains on this trip, we had again been granted carriage C1, perhaps a grim reminder of our market research status (it's a social grading demographic). Amazingly, everyone in the carriage was white, while the other five carriages contained no white people at all. An impossible coincidence, but how on earth this could be successfully prearranged is yet another mystery of the Indian Railways. The six-hour journey spent discussing the vagaries of the Indian experience with a retired English couple was most refreshing, and listening to the British 'Jewel of the Raj (sic) adventure package' tourists behind us moaning about the food, hotels and weather was not. On the way up we spotted a number of Himachal pheasants, wonderfully strange birds and rare as the proverbial hen's teeth. We left our countrymen and headed off to check in with the Royal Family of Jubbal at their residence, Woodville Palace. Originally constructed at the behest of the Commander-in-Chief of the Imperial Army, it is now a terrific country hotel, and the grandson of the Raj of Jubbal apparently lives on the first floor. We had booked this wonderful hotel principally because Julie had seen Dirty Den from Eastenders talking about his stay there on a television programme. We were disappointed that despite quite persistent enquiry and a good look at the guest book, no one had any recollection of his stay.

The morning brought a clarity and calmness long forgotten. In Shimla there is no hassle, a place you can go for a walk without even being looked at. Since landing in Delhi we'd noticed how much less interesting we were as white people, but in Shimla it is even possible to go into a shop, pick something up and the put it down again WITHOUT COMMENT. Shimla was truly wonderful, the shambles of hotels and houses clinging to the mountainside like scree were a far cry from the overpopulated, over-commercialised hill station that the guidebooks and backpackers would have us believe. Yes, it is a tatty ersatz holiday town with plenty of crummy hotels and a screaming child or two, but it is Shimla for Heaven's sake, it is popular for good reason. Our first sight of the Himalayas from the Jakhu Hanuman Temple was momentous and we started planning a trip further north for next year. Watching a tree creeper twitching on the bark of an altitudinous deodar tree made us again think of home, in fact Shimla made us think of home in so many ways, coming here with only a few days left in the country was the right thing to do. As we wandered around the steep lanes from one end of the ridge to the other, the countless public schools started knocking off for the day and hordes of small uniformed people filled the mountain. A group of boys approached us and asked where we were from and if we could speak Hindi. I replied that we could not.

"*Chutia gauri*" a short round boy ventured to me, giggling and making eyes to his friends.

We walked on and he strode confidently behind us, "*Bhen chod*", he asserted, sounding very pleased with himself.

I turned on him unheralded. "When I said I couldn't speak Hindi, I lied." I lied.

He stopped in his tracks and his eyes dilated in fear as his friends dispersed like pigeons. "A…a…a…" was all he could manage.

"I know that you first called me a white bastard and then a…well, it's not repeatable, but it relates to my sister and I."

The colour drained from his face and he almost fell to his knees, here was an inexperienced boy with newfound cockiness, which looked now to be very short-lived. "Sir, please I am very sorry, I did not mean anything, I will not do this again sir, please…"

I shrugged a half-smile and slowly nodded my head before closing my eyes, the symbol for him to turn and run. Where I grew up, it wasn't the posh kids who called you names, it was me.

The return journey to Kalka was a six-hour hell. We were again in carriage C1, and I was pleased to see there were only a couple of other white people. What I didn't know at that time was that our tiny-seat neighbours were to be the upper class disgrace of the nation. As they boarded, the grandparents, parents and two kids ordered their *coolies* to leave their luggage where it would be most obstructive and sent them away with a nonchalant wave of a twenty-rupee note thrown backwards on to the floor. They did not acknowledge us at all, despite our smiles and helping hands, but arranged their two young girls on their knees so that their feet swung to kick our shins. The grandparents grumbled on the adjacent seats, occupying one each. After half an hour we had been trodden on, kicked, dribbled on and been used as dirty hand wipes for these two whinging brats while their parents made no apology or even acknowledgement of the upset. I'd never seen anything like this before in India, of course it's commonplace in Britain, but Indians are always polite and always in control of their kids. I finally asked them to try to control the children and was met with steely glances. Next the top of the range video camera came out and the children charged up and down with it presumably filming a blur of the floor while the parents closed their eyes. The camera was finally abandoned on the floor in response to the arrival of sweets, crisps, sandwiches and drinks. As the first sweet rapper went out of the window to dance around the mountains forever, I winced. The second one saw me tensing my fingers violently on the chair. When the bumper sized crisp packet was emptied and handed to the father I felt some hope, but as he tossed it wantonly out into the wilderness I snapped.

"Excuse me. Any more litter you have, give it to me and I will put it in a bin when we arrive. Please do not throw anything else out of the window."

He looked my way momentarily and said nothing. I could see his wife being given an empty plastic water bottle by her child and as she tried to throw it out of the window I slid the window shut in protest. As if I just wasn't there, she opened the window from the middle and stood to throw the bottle out. I jumped up and grabbed it from her hands, crushed it and placed it under my seat. She looked a little embarrassed but minutes later leant across to the window on the other side of the carriage where her parents were sat and threw out another huge crisp packet. I lost it.

"You animal," I shouted in her face, "You disgust me. This is a National park for God's sake, have you no morals? You are actively encouraging your children to ruin the countryside. Is there not enough litter everywhere for you? Do you want to live like pigs?"

As you might imagine, after this outburst things became a little tense in the carriage. But the family did not respond verbally at all, they just looked right through me. For them, I simply didn't exist, despite my direct protestations. I suppose this is how middle to upper class Indians deal with the social inequalities around them; and to these people, by raising my voice, I had become just another lower caste scumbag to ignore. Really middle class India, this just isn't good enough.

Changing trains and pulling back into Chandigarh was a great relief. I was wound up so tight by such flippant littering that I requested we head out for a few ripeners and unstiffeners. India were playing Pakistan in only the second test series they had consented to in fifteen years, and the first in India, and understandably, Chandigarh had been chosen as the venue for the first test, given its proximity to Pakistan. As we walked into sector 17, the place was mobbed. There were signs all over saying 'Welcome to our cousins from Pakistan', a heart-warming sentiment considering the total resentment I had experienced toward Pakistan everywhere else in India. And they meant it, it wasn't just the shops trying to cash in new money for forbidden goods, though the tills in the film and music shops were ringing off the wall. In the main square we saw Abdul Razzaq being sardined by Indian cricket fans patting him on the back, shaking his hand, proving (rather sadly) that greatness transcends prejudice. We saw one man walking through the six-foot sea of turbans wearing a 'Proud to be a Pakistani' cricket T-shirt, bringing smiles and handshakes at every turn, very brave, possibly a result of some Dutch Courage imparted from one of the local bars that were very popular with the Pakistanis that night.

On the train back to Delhi it occurred to me that I knew the origin of almost all of the mobile ringtones around me. As phones rang throughout the carriage, the hit songs from our eighteen-month stay beeped, tinkled and tolled to my subconscious; evocative reminders of all the things we'd done and the places we'd been. I realised I wouldn't be able to do the same thing back in England, I'd missed so much time I wouldn't have a clue, and I realised I had become much more of a part of my surroundings than I'd been letting on to myself. I contemplated how much more of the culture and language had actually crept in and laughed to think how often I found myself saying *achcha* and *teekay* instead of ok, how neither Julie or I could now avoid the head-wobble while agreeing with people, and I wondered how I would fare back home saying *bolo* or 'tell me' rather than hello when picking up the phone.

Before leaving India there was one more thing we had to do, one obvious outstanding sight to see before we headed back home. You know what it is. It's the big white thing that everyone associates with India, the 'ultimate monument to love', the 'eighth wonder of the world' and every other cliché you can think of. To be honest I really wasn't keen. Everyone always bangs on about how great it is, saying 'don't underestimate it' and 'you can't believe it when you see it' but this just makes me want to hate it more, and I was concerned that I'd do what I so often do when faced with prescribed symbols of greatness and wonder, and just rubbish it to be contrary. We toyed with the idea of not going, pretending to go but not really bothering, its much easier, I've done it many times before.

"So you've been to Egypt?"
"Yes."
"What did you think of the pyramids?"
"Oh they were wonderful."
It's that easy.

But we had booked a day return ticket and couldn't really think of a good reason not to use it, so we went. Arriving on the train from Delhi, we had expected to run the gauntlet of beggars as we approached, but like many other hassle hotspots (Chandni Chowk, Colaba, Anjuna, Varanasi), Agra has been cleaned up, the limbless, diseased and inflicted turfed out to aid tourism. They don't even allow mobile phones inside the Taj grounds, so Julie and I deposited ours with two men who sat behind a table laid out with hundreds of phones, all turned on, and all ringing. I asked the man how he coped with the din.

"I must be happy, it is music, I sing along see, La La La La La La …"

One's first view of the Great White Mausoleum is pretty spine chilling. It's such a familiar image that it feels something like coming home. But I have to say it soon wears off. After obligatory photographs and a little poking about, we were wondering what else to do. We thought we'd better go in, so we gave our shoes to the attendant and stepped inside. WHAM - the lingering smell of three million smelly feet every year punched us full in the face. Yes, the Taj Mahal looks fantastic but it smells like a cheese shop.

Wondering what else to do we stretched out on the lawns and I considered what was left to see of India. A great deal, no doubt. It would be easy to say that we had seen very little, and in some respects that would be true, but we had seen *something* certainly; for a start we had seen pretty much every bird in the book, every bird that is apart from the hoopoe, a bird which is apparently common but despite great efforts, had evaded us. As I was contemplating how looking for birds is a great metaphor for looking at foreign lands, how seeking them out is very much like seeking out the *real* aspects of any unfamiliar place, and as I was thinking that being eluded by often the most common species is much akin to missing the point in new cultural exchanges, two hoopoes swooped in and landed on the lawn in front of us. Two magnificent hoopoes to round off the entire set. Two plump orange, black and white woodpeckers scurrying about the turf as if on tiny bicycles, pecking at the ground by the Taj Mahal.

I clapped, unable to do anything else.

****

The final night of our Indian odyssey would be spent at a wedding between two people we'd never met. This was not really the finale I was looking for, but then again, when I considered it, I didn't really know what I wanted. We were exhausted after our gruelling railway tour and anxious as hell about finally going home. But we'd agreed, and that was that, and the more I thought about it, the more agreeable it seemed, we'd be driven to the wedding with our luggage in the car and could go directly to the airport from there. And we wouldn't have to think about anything.

The married couple looked exhausted. I gathered they had been married some time earlier, and that the celebrations had been going on for roughly seven days. The celebration was everything one would expect from two wealthy Punjabi families in Delhi: grand, glitzy and centred around the consumption of enormous amounts of food. The party looked like it would go long into the night, but we couldn't really engage with it, we couldn't keep our eyes off the watch. When the hour finally arrived we said our farewells. It was surprisingly emotional for us, leaving people who, apart from Shanti, we had known less than two weeks. But now these people represented our India and everything we were leaving. We hugged them sincerely and promised to return.

Chapter Eighteen

"Don't you think these suitcases are a little heavy?" the check-in clerk enquired with a wry smile.
"Err, I guess so, its hard to leave anything behind, India is such a wonderful place." I creeped.
"Do you have any hand luggage?" the clerk asked, still smiling.
"Only this." I lied trying to look sincere while holding up Julie's handbag and concealing the two cabin cases and laptop computer at my feet.
"OK, no charge."

India, oh how I love India.

We went through to the airport lounge and Julie went off to get a cup of tea while I sat and tried not to reminisce too much, but as I looked around, I grew increasingly sad. How I would miss this comfortable chaos, this fervent atmosphere of feeling. This dirty, smelly, noisy, sweaty, overcrowded and dangerous country, this mysterious and superstitious nation, where everything moves too fast, too slow or too erratically and where everyone is too poor, too rich or too indifferent to notice. How I would miss this India.

I thought of our achievements and felt better. I thought about everything we had to take back with us, the 'Eastern Sense' that we'd tried so hard to encompass. And I thought of the book I'd written and the new life ahead, a life without work, without the pacifier and without the same rules, and I was proud to have been changed by India.

And we would return, there would be holidays...the phone rang. It was Avinash. He must be calling to say goodbye.
"Hello."
"Mick, its Avinash, how are you?"
"Yes, fine, we're at the airport. Not long now..."
"Mick, listen, I know its very late in the day, but I wanted to catch you before you left. I have a proposition for you..."
"Hang on there a minute, its not related to work is it? I told you I have

made my decision and I'm quitting the research game. I've decided and I'm not changing my mind. I have to make the break once and for all, and if I don't make it now I never will. Whatever it is, the answer's no."

"Mick, I know, I know. This is not really work. Well it is, but it's probably more like fun for you, just hear me out ok? We just secured a new client for research and they want to station someone here in India, things will be run from the Delhi office, and for three months they want someone to show their chap the ropes, show him round Delhi, help get him acclimatised, and we want someone to, shall we say 'shield' the client from our staff until they get to know how things really work. The truth is we don't want someone coming over, not liking the way we do things and then pulling the project, we need some enthusiastic hand holding, in short, we need you."

I was silent.

"Mick? Are you still there?"
"I'm here, but I've given you my decision already."
"Mick, at least think about it. You don't have to start for another month, so you can go home for a while, and its only for three months and we'll make it worth your while, you can bring Julie too, and we've just rented a fantastic flat in downtown Delhi, fully kitted out with all mod cons, you'd just have to move in, it would be like a luxury holiday…what do you say, will you think about it?"
"…well…err, when would I have to give you an answer?"
"Oh, not until sometime next week I guess, go home, have a think and let me know, but call me anytime to discuss OK?"

"OK, I'll think about it. I have to admit, it does sound interesting…" I looked up to see Julie coming back with the teas, "Oh, Avinash, just one thing, this flat in Delhi with all mod cons…does it have a bath?"